RADIO AMATEUR OPERATING HANDBOOK

by Marshall Lincoln
W7DQS

EDITORS and ENGINEERS
Division HOWARD W. SAMS & CO., INC.
INDIANAPOLIS, INDIANA 46268

FIRST EDITION

FIRST PRINTING—1972

International Standard Book Number: 0-672-24028-9
Library of Congress Catalog Card Number: 77-185177

Preface

The fascinating field of amateur radio is composed of two basic functions: experimenting and communicating. Virtually all ham activities are related to these basic functions. The experimenting function consists of building, testing, and modifying equipment. Putting this equipment on the air for a useful purpose encompasses the communicating function. No matter how fancy an amateur station layout may be, it is not effectively utilized until the operator has mastered the essentials of communicating his thoughts by radio.

This book deals primarily with the communicating function, and its aim is to help you understand and master the methods which have been developed through years of experience by many thousands of other amateur radio operators. Successful communication techniques are often taken for granted, but they need to be considered just as carefully as the technical aspect of radio. The ham who devotes as much attention to his operating methods as he does to his technical achievements is bound to accomplish the most with his hobby and derive the most satisfaction from it. Just as an antenna can be tuned for maximum transmission of radio energy, so also can an operator's methods be tuned for maximum transmission of thoughts or messages. Today's extremely crowded band conditions make it more important than ever before to use the best operating techniques possible.

Many amateur radio operators have assisted in making this book possible. It is impossible to list them all and dangerous to list only a few, but it is also only fair to list certain ones who have been especially helpful to me and have taught me many things, which I now try to pass along to others. Among those operators who helped sow the seeds for this book, most of them without knowing it at the time, are W9CXL, W1BDI, W1NJM, W9FZW, W9FWH, W9TT, WA9BHV, W7JCH, and WA6AWD. Actually, though, every amateur I ever met has helped in the writing of this book, for I have learned something from each one.

MARSHALL LINCOLN, W7DQS

Contents

Chapter 1

Making Yourself Understood

To see just how much can be accomplished through efficient operating techniques, let's take a quick look at a couple of commercial-type communications operations and see how they do it. Two prime examples of busy, no-nonsense radio applications are to be found on the aircraft and police frequencies.

It is not intended that hams should imitate these types of communications, but it is also certainly true that we hams can learn much from them. In both aircraft and police radio operations, there are many persons on the air, in rapid succession, and the content of their messages often carries great urgency. Yet, through efficient operating techniques, it is possible for all these persons to exchange vital information quickly and clearly.

How do they do it? Let's look at aircraft radio first for some examples. An important item is the use of frequencies having definite designations for definite purposes. Pilots have a wide choice of operating frequencies, just as hams do, but each one has a specific purpose. A pilot approaching Chicago for a landing would call the control tower on a frequency designated for that purpose. Another pilot might be just a mile away, calling for the latest weather reports from Kansas City, but the two pilots' transmissions would not interfere, and there would be no confused tuning to find the station each wanted to talk to, for the frequencies for each are known to everyone involved.

Pilots also use certain terminology that allows them to say a lot in just a few words. Since this terminology is *standard* throughout the field of aviation, everyone understands just what is meant. For example, a pilot may report he is "eastbound 10 north of Peoria leaving five for three." That transmission takes only a couple of seconds, but it carries this complete message: he is eastbound, is 10 miles north of the city of Peoria, and is beginning to descend from five thousand feet altitude and will level off at three thousand feet altitude. Pilots also make use of a phonetic alphabet, but they use it sparingly, only when needed to avoid misunderstanding of important information.

Policemen go to even greater lengths to condense their transmissions into rapid bursts of information. The various code numbers used in police radio work are not so much intended for secrecy as they are intended to convey a large amount of information in a very short space of time. Each man gets his message across quickly and gets off the air to give the next officer a chance.

Another important aspect of both aircraft and police radio operation is the fact that in both cases, *listening* carefully is just as important—often it's more important—than transmitting. These frequencies are too busy for the men involved to be asking a lot of questions to find out what everyone else is doing. This information is acquired by everyone listening alertly.

When an airport control tower operator reports the location of a certain aircraft approaching the field, there often are dozens of pilots in the same area who hear this transmission and store this information in their minds. Seldom does anyone need to take up time on the busy frequency to ask for information about the plane being reported, for everyone has been attentively *listening*.

Policemen do likewise as they hear other cars in their area being dispatched and patrolmen reporting to headquarters on their activities. An important key to successful use of radio is the ability to *listen* well.

Hams do not usually need to adopt the exact techniques of pilots and policemen (although some of our DX contests may sound like a mid-air collision or a roaring shootout!), but we can learn much from them and apply it to our own communications. For a practical lesson in efficient ham radio operating, let's take a critical look at the use of a phonetic alphabet.

USING PHONETICS

The sole purpose of phonetics is to be able to speak letters so they can be clearly understood in voice communications. A phonetic alphabet is a set of familiar-sounding words which may be used in place of letters when needed if noise or interference, a weak signal, or the similarity in sounds of some letters prevents the human voice from being clearly understood by the receiving operator.

The object is to be *understood,* not to be cute. Because of the informal nature of much of the communicating done by amateur radio, some hams have adopted special phonetics which have a humorous touch. Although these can make conversation openers, or get a laugh from someone hearing them for the first time, they can become pretty silly at times, and actually are very much out of place in serious radio operating.

Another misuse of phonetics is the *overuse* of them—for example, the habit of spelling all call letters, names, and cities phonetically, even when the receiving operator

Modern amateur radio stations include a wide variety of both commercially built and homemade gear, assembled to suit individual operator's tastes and interests, and giving capability to operate many transmission modes. By following good operating techniques and procedures, amateurs can make best use of this equipment, regardless of operating preferences.

is having no trouble hearing what is said. Phonetics are intended to improve clarity when noise or interference make things difficult for the receiving operator, and to avoid confusion over letters that sound nearly alike. So, obviously, there is no need at all for phonetics except for these purposes. If it becomes necessary to spell out every word you say, then conditions are too bad for any useful communications by voice.

Some operators have the habit of consistently using phonetics in giving their calls and their first names, even though receiving conditions are good, then continue for a lengthy radio contact without ever using phonetics again! Often they didn't need to use phonetics in the first place, but did so out of habit, without thinking of the real reason for having a phonetic alphabet.

Even when conditions are bad and phonetics are needed, they usually are needed only for the essential details of the transmission, with other parts of the transmission understandable because of its general context. For example, in giving call letters phonetically, it generally is not necessary to give the prefix and the call district number phonetically—just giving the call letter suffix generally gets the message across quite well. The call W7DQS, for example, would be spoken: "W 7 DAVID QUEEN SUSAN." In giving the name, it would be transmitted: "The handle is Marsh . . . I spell MARY ADAM ROBERT SUSAN HENRY."

In the case of very common names, it is usually not necessary to use phonetics unless radio conditions are especially bad. Names like Bob, Bill, Joe, Sam, and Don, for example, are so common that they are usually easily understood.

Several years ago, John Troster, W6ISQ, wrote a very humorous satire in *QST* magazine (August, 1962) on the excessive use of phonetics by operators who feel compelled to spell out nearly everything phonetically. The title of his article, "All Otto King On The Queen Roger Mary, Charlie," gives an insight into the effective way he lampooned this silly practice.

As to what words to use for phonetics, you can easily get a lively fight started at a radio club meeting on this topic. Most operators who have been on the air for many years seem to prefer simple, home-grown words such as "Able, Baker, Charlie, Dog, Easy, Fox, George," etc. for their phonetic alphabet.

Younger operators, or those exposed to military communications, have generally taken up the newer, sterile-sounding phonetic alphabet which goes "Alpha, Bravo, Charlie, Delta, Echo," etc.

The latter alphabet is officially called the "ICAO" alphabet, for it was adopted by the International Civil Aviation Organization for use around the world by pilots. Supposedly it is better because it can be understood when spoken by persons speaking different languages or dialects. This may be true when radio conditions are good, but it seems doubtful that anyone could clearly understand the word "Golf" used for "G," for example, during a burst of radio noise. Good, old "George" seems much better suited to use during periods of poor radio reception. Also, the use of "Quebec," which the ICAO boys say should be pronounced "Kay-Beck," to indicate the letter "Q" often throws a receiving operator into a temporary quandary. "Queen" seems much more suitable, and actually is generally used by pilots—even those pilots who use the rest of the ICAO alphabet pretty much as it was written.

Many years ago, the American Radio Relay League published its "ARRL Phonetic Alphabet" as a suggested list for hams to use to provide a reasonable amount of standardization. This list, consisting of familiar, easy-to-remember everyday words like "Adam, Baker, Charlie, David, Edward," etc. was patterned after the phonetics used successfully by military operators in World War II.

RADIO SHORTHAND

Phonetics are used to improve *clarity* in radio transmissions, while a number of verbal "shorthand" expressions are used for the sake of *brevity*.

Just as pilots and policemen, used in the examples at the beginning of this chapter, use shortened terminology and specialized codes to save time, so also can hams use specialized words and phrases in radio communications.

The use of time-savers originated on cw, but as voice operating became more widely-used in radio, special shortcuts were derived for use on phone.

These shortcuts take several forms, and all of them are useful in everyday ham operating. Everyone knows of the "Q code" devised for use on cw, for it is included on ham license exams. The meaning of each Q signal is precise, and included in a listing in many radio books (including the Appendix of this one).

But, many operators are not fully aware of the precise meanings of some other pieces of radio shorthand, called "prowords," meaning "procedure words" or "procedure phrases."

These prowords are an important stock-in-trade for the pros—the commercial operators. They can be equally useful to amateur radio operators, when used properly. The cw counterparts of prowords are "prosigns," meaning procedure signals.

Let's start with a study of prowords used on phone, and follow with the prosigns used on cw. You will soon see that their use can make anyone's operating more efficient. They are all listed in the Appendix of this book for reference, but their proper usage will be explained here.

PROWORD USAGE

The everyday expression "go ahead" is a proword, although it has become such a commonly heard one that most of us do not think of it as such. "Go ahead" is a standardized way to say "I'm through transmitting for the moment, so you begin transmitting, and I'll listen." That's quite a mouthful, but you can get the entire thought across with just two words.

"Over" is another proword with the same meaning. Both "go ahead" and "over" are standardized *invitations to transmit*.

"Stand by" is another proword, and it is important to recognize its precise meaning if you are to avoid needless confusion on the air. "Stand by" means, in effect, "please wait on this frequency, but *do not transmit*." It is a definite, although polite, *command* not to transmit. Often it is OK to acknowledge it with the briefest possible transmission, but after that, the well-disciplined operator will maintain complete radio silence while carefully monitoring the frequency, prepared to transmit, but not yet transmitting.

"Roger" is another common proword—one that is very commonly misused. "Roger" is often used loosely to indicate "yes," "okay," "I agree," "that's correct," and other related meanings, but it actually means *none of these!* The correct, precise meaning of "roger" is simply: "I understood your entire transmission." It means nothing more, and more importantly, *it means nothing less.*

"Roger" is a precise acknowledgment to another operator that you understood *everything* he said. He then knows he does not need to repeat anything from that transmission. "Roger" does not necessarily mean that you agree with what was said, or that you will do what might have been asked of you in the transmission.

A proword you may occasionally hear from an older operator with military experience is "wilco." Most newer operators never heard of it, but it saw wide use in World War II communications. Its meaning is "I *will* comply." Its only use is in cases when an operator gives instructions for some action to a second operator.

The simple reply "wilco" means the second operator understood the instructions and will carry them out. Operators who used "wilco" in the military services point out that to say "roger, wilco" is redundant; for if you are agreeing to comply with the instructions, then you obviously understood them, so in this case, "roger" is unnecessary.

On cw, the letter "C" means "yes," so some military operators use the phonetic "Charlie" to mean "yes." Sometimes this becomes "Charlie Charlie" for emphasis, or even (heaven forbid!) "Charlie Brown." The unfortunate tendency of some hams to ape some of the strange "mod-sounding" military communications terminology means you'll occasionally hear these words on the ham bands. They all boil down to a simple "yes."

"Affirmative" also is used by many operators on phone to mean "yes," and "negative" to mean "no." Some old timers shudder at this usage of long words in place of short ones, but younger operators seem to prefer the brisk, business-like sound of "affirmative" and "negative." It must be admitted these longer words can be understood better in noisy radio conditions than "yes" and "no," which many of us tend to slur badly into mere grunts.

"Break" is another often-misused proword. Properly used, it indicates that a station wants to "break in" to an existing communication. The "breaking station" is the newcomer who wants to join the conversation, or who needs brief use of the frequency and wants to ask those using the frequency to stand by. The most successful, and easily understood way to use "break" is to say "break, this is W7DQS," or "break, this is W7DQS with important traffic for W9EL," for example. Then *stand by,* and wait to be called by a station already using the frequency.

Only one "break" is necessary, except for a very specialized use of "break break" and "break break break" to be explained shortly. Some operators say "break" when they are cutting short their own transmission to allow another station to transmit. They may ask a question, then say "break," without signing their call, to indicate that the operator of whom the question was asked should go ahead. This use of "break" is incorrect, and can be confusing. Better to just say "go ahead" to the station you wish to transmit. (And to spare everyone a lot of unnecessary verbiage on the air, remember it's only necessary to sign your call once every 10 minutes— *see* Chapter 2.)

Some operators say "break" (or send "BK" on cw) when they resume transmitting after the second station has responded to their first "break." This use also is incorrect and unnecessary.

Some phone traffic operators will say "break for check" when they stop in the middle of a message they are transmitting to determine if the receiving operator is copying okay. Also, "break" is used to separate the preamble, address, text, and signature of a formal message sent on phone. These are legitimate specialized uses of "break" to be covered in Chapter 7 on Message Handling.

There is another specialized use for "break," and even an approved use of "break break" and even "break break break," which normally are frowned on by serious operators.

These uses were started by a group of West Coast hams who organized what has become a widely-recognized full-time public service net. It's called WCARS, meaning *West Coast Amateur Radio Service*. Members monitor 7255 kHz and have established an enviable record for assistance to other hams and to the general public in a variety of ways.

Since its inception a number of years ago, WCARS has been joined by MIDCARS *(Midwest Amateur Radio Service)* and ECARS *(East Coast Amateur Radio Service)*, so that the entire nation is spanned by member operators daily.

To provide for efficient flexible operation, these 40-meter net members have established a series of special proword usages to fit their type of operation.

In these nets, "break" is used by a station desiring to join a net session with routine or informal traffic; "break break" is used by a station with priority or urgent traffic, and "break break break" is used for emergencies only.

However, in other amateur work apart from these particular nets, use of "break break" or "break break break" is generally considered to be the mark of a poor operator.

Further explanation of prowords used by these nets are explained in Chapter 7.

"Clear" is a proword that has crept into phone operating, without much official recognition—at least yet. The general meaning conveyed is that the operator has completed his transmissions, but will monitor the frequency for awhile in case anyone wants to call him.

This proword apparently has been picked up from various commercial services, such as police, taxicabs, public utility companies, and the like. Since these services do not use identical procedures, you will find hams' usage of "clear" is not always precise. In time this proword probably will acquire a more universally understood meaning.

"Out" means the sending station operator will not be available for calls, at least for the moment. It *does not mean* that the operator is ending his transmission, but will be standing by for additional calls. Occasionally an operator will use "out" in this sense, but he really should be using "clear."

There are a few expressions frequently heard on the air which are meaningless or blatently incorrect. One of these is "QRX one." In the first place, this one is off limits because Q signals are intended for cw (or rtty) only; there's no point in using them on phone, and serious operators fly into a screaming snit at anyone who does use them on phone. In the second place, "QRX" used to mean "stand by" or "wait," but this meaning was dropped from the Q-signal list many years ago. "QRX" now means "I will call you again at . . . (time)."

"Over and out" is a contradiction, because it means "go ahead, but I'm not going to listen." Although there probably are several people you'd like to say this to (and all of us sometimes practice this procedure of closing our ears without letting on), still it is a pointless thing to say in serious radio operating.

"HI" as everyone knows is a cw expression used to denote laughter. It is frequently used to allow a ham to laugh at his own jokes on the air. When sent by a skilled operator, the dits spurt out so they sound like a sprightly chuckle.

But, thanks to the availability of a microphone to pick up the sound of the human voice, there is obviously no need to use "hi" on phone to indicate laughter. (The often-heard "hi hi" ranks with the pointlessly plural "73s" as an outrageously ignorant expression.) To say "hi" on phone when you wish to indicate you are amused about something is ridiculous. Just do the most normal thing—*laugh.* The receiving operator will recognize a laugh when he hears it. And if you can't laugh convincingly on the air, then the joke wasn't really very funny after all, was it?

10 CODE

The usage of a new radio "language" is creeping into ham radio, mostly in the operation of hams who use f-m on 6 and 2 meters and on the 450-MHz band. This is the "10 Code," which you may first have encountered on some of the TV "cops and robbers" programs.

Good operating procedures go hand in glove with good equipment to extend the modern amateur radio operator's ability in all forms of operating, which may include traffic handling, DX and certificate chasing, contests, and informal rag chewing.

Various versions of a 10 Code have been used by police departments for many years, and other radio users in commercial and business-oriented fields have adopted various 10-Code signals for their own purposes.

With so many forms of radio communications in everyday use nowadays, just about everyone encounters someone using a 10 Code eventually, so it was only natural that its usage has begun to creep into ham radio. Amateur f-m operation, in most parts of the country, has started out using a more brisk operating style than the easy-going conversational style that has been common on the "low bands," and this faster pace of operating by f-m hams has made it inevitable that some of them would begin using a 10 Code.

Also, vhf and uhf ham bands become pretty crowded in large cities, with so many operators crowded into each other's operating ranges, so use of a 10 Code has helped ease this congestion a bit by providing a means of speeding up transmissions and getting more information transmitted in a shorter space of time. Also, a 10 Code can be helpful when used by a group of hams engaged in an emergency situation, by providing a commonly understood way to exchange information quickly, and to keep track of each other's activities without excessively long transmissions.

Generally, hams in a given city will pick up a few of the 10-Code meanings used by police or commercial users in their area, and begin using these on the ham bands. However, police and commercial interests do not have a universally understood 10 Code themselves—code meanings differ in various parts of the country. So, ham 10-Code meanings also differ from city to city.

To provide a list that is accurate wherever you live or travel is impossible, but the list included in the Appendix is indicative of typical ham usage of a 10 Code, based on a sampling from various areas.

PROSIGN USAGE

Prosigns are the cw equivalent of prowords, and they make it possible to accomplish a surprising amount of communicating in a short time. Hams who turn up their noses at cw because it involves spelling everything out by hand obviously have not discovered that there are so many cases in which use of cw "shorthand" allows transmissions to be made even faster than by voice!

A complete list of cw prosigns appears in the Appendix, but their meaning and use will be explained here.

The simplest prosign is "CQ." We all know it means "calling any amateur station." Hams adopted it years ago from commercial operators, who used it to mean "attention all stations."

"K" is another common prosign. It corresponds to "go ahead" or "over" on phone, and is simply the cw invitation to transmit.

When calling a specific station, and desiring that no one else interrupt, a ham may send "$\overline{\text{KN}}$" instead of merely "K" to emphasize that no "breakers" are wanted. (The line over the letters indicates they are run together into one code character in transmission.) This may seem like a discourteous attitude for hams to have, and it is not needed very often, but occasionally, such as in an emergency situation, or while completing a DX exchange or a contest contact, an operator may use "KN" to prevent needless interruptions. It also may be used after a directional CQ call. For example, an operator might call "CQ ARIZ CQ ARIZ CQ ARIZ DE WA9YCA KN." This would emphasize that calls are wanted only from Arizona.

The prosign "AR," which also is sent with the code characters run together as if it were one letter, indicates the end of a transmission. It is used in message handling to indicate the end of a formal message (more about that in Chapter 7), and it is used following a call to a station with which contact has not yet been established.

For example, W7DQS might call "W9BS DE W7DQS "AR." After contact was established, the "AR" would be dropped and "K" used instead as an invitation to transmit.

"AR" also is used by some operators to indicate the sending station has completed what he has to say for the moment and is preparing to sign over to the receiving station. For example, if W7DQS is sending to WA7BIY, as he completed his transmission he would send " . . . AR WA7BIY DE W7DQS K." This gives WA7BIY a brief warning that it will soon be his turn to transmit.

"BK" is sometimes heard on the ham bands, apparently meaning "break," during an on-the-air visit, but as pointed out in the portion of this chapter dealing with prowords, careful operators see no need for this one, so it is generally considered taboo when used this way. They feel the simple "K" carries the correct meaning, so there is no need for "BK."

"BK" may be used correctly by a cw operator desiring to join a radio contact already in progress, or by a station wishing to check into a cw net, just as a phone operator would say "break."

A similar prosign is "BT," again with the characters run together as if they were one letter. This is often thought of as a dash, such as you would write to separate two thoughts. Most cw operators do not use periods and commas in ordinary conversation on the air, but use "BT" instead to separate sentences or phrases as they send. "BT" also is used to separate the principal parts of a formal message —this also is covered in Chapter 7.

"R" is used frequently on cw—and seems to be misused just as much as "roger" is on phone. This is probably to be expected, since they both mean the same thing!

As already explained, "roger" means "I understood (or copied) your entire transmission." And "R" on cw means the same thing. The old timers say that "R" came into use long ago, before the beginning of voice operation on radio. Perhaps it originally stood for "received." It meant they had copied the entire transmission—no repeats were necessary.

With the advent of phone operation on radio, operators picked the phonetic word "roger" to indicate "R" when operating by voice. Just as on voice, however, "R" does *not* mean "yes," "okay," "I agree," or anything of the kind. "R" sometimes also is used on cw to indicate the decimal point in a number. So, "123.45" may be sent "123R45."

"$\overline{\text{AS}}$," sent as a single letter, means "wait" or "stand by." Once upon a time, "QRX" was used to convey this meaning, and still is by some hams who haven't realized that the meaning of QRX was changed many years ago to "I will call you again at . . . . " There is no Q signal meaning "wait or "stand by" now. Use "$\overline{\text{AS}}$" to convey this meaning.

"$\overline{\text{SK}}$" means "I am finished with all I have to say, but I will listen for you if you have anything more to say." "$\overline{\text{SK}}$" is used at the conclusion of your final transmission. It has the same general meaning as "clear" on phone, and generally is sent before you sign the necessary call letters.

For example, a final transmission by W7DQS at the completion of a contact with W9QLW would be: " . . . 73 $\overline{\text{SK}}$ W9QLW DE W7DQS K."

It is said that this prosign originated from the American Morse code, which is somewhat different from International or Continental Morse Code. Old-time telegraph operators, who used American Morse Code (known also as "land-line Morse"), had a code signal "30" which meant the end of a message. When sent rapidly in American

Morse, "30" sounds very much like "$\overline{\text{SK}}$" in the International Morse Code used by radio operators.

Incidentally, "30" is still used on some newspapers at the end of a piece of copy as a signal to the typesetter that there will be no more "adds" (additional paragraphs) coming from the copy desk on that particular story. This custom has been carried over from the days when out-of-town news items were sent to newspapers by manual telegraph, and the telegrapher put "30" at the end of each dispatch he sent.

(The copy desk editor who handles incoming stories from distant cities is still called the "telegraph editor" on newspapers today, although this copy has moved by *Teletype* rather than manual telegraph for several decades. Customs such as these live long after the events which created them.)

"CL," which is sent as *two letters, not run together,* means "closing station." (Do not make the mistake of assuming that it is an abbreviation for the proword "clear" used on phone—they do not mean the same thing.) "CL" is sent at the very end of a final transmission, after signing the call letters, to indicate that the sending station is ceasing operation for the moment and will not answer any other calls. If K5MAT was closing down at the end of a contact with K7NHL, he would send "K7NHL DE K5MAT CL."

The letter "C" sent by itself means "yes," and "N" by itself means "no." Be careful though; don't try to emphasize a "no" by sending "N" twice—the receiving operator may think you sent "C!"

There are some additional prosigns with specialized purposes for message handling and net operation; these are covered in Chapter 7. You can see from the preceding description that intelligent use of prosigns on cw can help move a radio contact along quite briskly.

CW ABBREVIATIONS

Just like abbreviations used in written notes, these are merely shortened forms of words, used to save time. There is no truly official list of these for ham use, but quite a number have been informally adopted by most operators in standard form. A list of these is in the Appendix.

The good cw operator uses abbreviations extensively in informal transmissions. With skilled use of prosigns and generally accepted abbreviations, a conversation can move along quite rapidly. Abbreviations are *not* used in the text of formal messages, unless placed there by the originator of the message, because they could, if misunderstood, cause the message to be distorted.

A very extensive list of abbreviations, called the Phillips Code, was used in the heyday of the land-line telegraph circuits to speed up messages. The words in the Phillips Code are much the same as the abbreviations used by hams, but since many of them have not been widely adopted by hams, they may cause some confusion if you set out to use them. You may hear some old time telegraphers who also are hams using the Phillips Code on the air, or you may want to experiment with it with a friend to see how fast you can send to each other. For this reason, a partial list of the Phillips Code is included in the Appendix.

Q SIGNALS

Q signals are a Godsend to the cw operator, for they enable him to shorten his transmissions tremendously, and still be clearly understood.

A standard list of Q signals is used by cw operators throughout the world—this includes commercial operators as well as hams. The complete list is quite long (much longer than many hams realize) and includes many highly specialized messages used by ships at sea, aircraft, and commercial traffic handling stations.

For ham usage, a considerably shortened list is used, but in most cases, each signal has the same meaning as in commercial service. So, when you use Q signals properly, you are truly following professional procedure in your radio operating. An important exception are the QN signals used in amateur cw traffic nets; these *do not* correspond to commercial QN signals, which were derived for aircraft radio usage.

A complete list of Q signals for ham use is in the Appendix, but a summary of proper usage will be given here. In almost all cases, a Q signal can be either a statement or a question. When intended as a statement, the Q signal is sent alone. When intended as a question, the Q signal is followed by a question mark.

So, "QRS" means "send slower" and "QRS?" means "shall I send slower?"

Q signals are not intended to be used as words in a sentence, because the Q signal by itself carries the meaning of a complete sentence. So, it is foolish to say "There is QSB on your signal." All you need to send is "QSB," which carries the full meaning "Your signal is fading."

It is equally silly to laboriously send "The QTH here is Ft. Wayne, because "QTH" means "My location is . . . . " So, "QTH FT WAYNE" is all that is needed to carry the full meaning of "My location is Ft. Wayne."

In a few cases, a Q signal needs to be followed with another brief item of information to make it complete. The previous example of the proper use of "QTH" is one of these cases. Other examples are signals such as "QRG" which means "Your frequency is . . . kHz." (For more on signal reporting, *see* Chapter 3.)

A couple of our Q signals have become adopted as words in ham slang, which may be what has caused many of us to fall into the bad habit of using many of the Q signals as words. The most common of these is "QSO," which has come to mean "conversation" or "contact" on the air. (An eyeball QSO is a face-to-face meeting of hams.)

Then there is "QSL," which means "I acknowledge receipt" when used on the air with reference to a message, but which has been in use almost since the origin of ham radio as the name for the cards which hams exchange by mail to confirm on-the-air contacts.

Another Q signal, occasionally used by old-time cw operators who despise the sound of a sloppy fist, is "QLF." This is a stinging sarcasm coming from a skilled cw operator

—it means "try sending with your left foot for awhile." It is not shown on "official" Q-signal lists, but it has existed for a long time in the mental files of old time telegraphers.

The meanings of various Q signals change slightly as time goes by, so if you compare a new Q-signal list with one printed several years ago, you will find a few discrepancies. These changes, just like the changes in editions of a dictionary from year to year, occur because of changing ways in which Q signals are used, and because of changes in the type of operating and the purpose for which Q signals are used by hams.

The Q-signal list in the Appendix of this book is the latest available at press time, including changes announced by the ARRL as this book was being written.

Chapter 2

Identification and Logging Requirements

Identifying the station on the air and filling out the log book after a contact, although both rather brief chores, probably have generated more discussion, confusion, and arguments among hams than any other operating activity. It didn't used to be that way, according to the old timers, but modern technical achievements and increasingly complex federal regulations have produced their toll.

These operating duties still are not really too complicated, but several important revisions have been made in the rules in recent years, creating striking changes in operating procedures that are contrary to much of the procedure that has been used by hams for many years. Some operators have quickly adapted to the new rules, while many others still stick with older procedures from the force of long-standing habit, and some operators use a combination of new and old procedures. Also, in some cases, FCC officials have issued contradictory statements on what is required in certain types of operating. Consequently, there exists quite a bit of uncertainty among hams regarding just what is the proper procedure.

Adding complication is the fact that different modes of communication have different and sometimes contradictory rules governing them. For example, what is legal on rtty is not legal on ssb in some cases, and what is sometimes required in home-station operation is not required in mobile-station operating.

Consequently, there is no wonder that there exists quite a lot of confusion and uncertainty among hams of just what is really required to make on-the-air identification and logging legal and proper.

This chapter will attempt to explain the rules on ID and logging as they stand as this is written. As time goes by, more changes probably will be made, just as they have in the past few years, so when in doubt, check an up-to-date copy of Part 97 of the FCC Rules and Regulations which govern amateur radio in the United States, or consult a reliable Ouija board.

Hams in other countries will have to look to their appropriate licensing authorities for the rules that apply to them, but perhaps they can understand from reading this chapter why hams in the United States seem to pursue strangely differing methods of identification and logging.

BASIC REQUIREMENTS

The basic rule that applies to all ham work, regardless of mode, is that all transmissions should be identified with your call letters and go in the log, according to FCC rules. This includes "CQ" and "QRZ?" calls, and even short test transmissions.

It is an unfortunate paradox that some hams seem to ignore this requirement for on-the-air identification and logging when making tests, yet *over*-identify when in contact with other stations on the air, actually going to a great deal of effort to give call letter identification much more often than is required.

In the ordinary informal "QSO" as hams have come to know it, each station is only required to identify *itself* at the beginning of a transmission or series of transmissions and every ten minutes afterwards, under the new rules. Then, at the end of its series of related transmissions, each station must identify itself and *one* other station with which it has been in contact.

This requirement has only been in effect a few years, and is contrary to procedures established through regulation and custom for a long time before, so there are many hams who still follow the older system, either from habit, or from lack of understanding of the new rules.

The older system required that identification be made every 10 minutes, and that each time this was done, each station had to identify itself and every other station with which it was in contact.

With most ham contacts consisting of only two or three stations at a time, this was no problem, and actually was pretty handy because it served to "keep score" so everybody knew who was talking to whom. However, as "round tables" came into popularity, especially among ssb operators, station identification according to the rules then in effect became a ponderous ceremony that had to be repeated every 10 minutes.

Another facet of the older identification system, which is still used by many operators, was that on phone, each operator was supposed to say "This is . . ." or "From . . ." just before giving his call letters. Full ID procedure under the older system would go like this: "W9FWH THIS IS W7DQS," or "W9FWH FROM W7DQS."

With the newer system, even "this is" and "from" are no longer required. In fact, the newer FCC regulations in force at the time this is being written do not even say which call sign should be given first, although it has long been the custom for the transmitting operator to give his own call last.

On cw and rtty, "DE" has been used for years to indicate "from" in station identification ("de" means "from" in French), but even that is not actually required anymore, although most operators still use it, mostly because operation would get awfully confusing without it.

So, you can see the new rules certainly allow much more freedom in on-the-air identification by cutting it to a bare minimum. However, even those operators who are fully acquainted with the newer requirements generally find that some additional identification of transmissions is needed to avoid confusion.

It is quite possible, on crowded bands, for an operator to identify legally, and yet for no one else to know for sure just whom he's talking to! He may be in range of two separate "QSOs" involving different groups of hams, with some members of each group uncertain just who is talking to whom.

Even the owner of a little sports car can operate amateur radio on the road, although finding a place to mount the equipment takes ingenuity, as shown here. Special regulations covering mobile logging requirements are more liberal than those for home-station logging to avoid needless distraction of a vehicle driver.

In a traffic net, this situation can get very confusing if only the bare-minimum legal requirements are followed in ID procedures. So, most active operators today are seeking a compromise in their operating habits between the excessively strict requirements that were on the books for many years, and the extremely loose requirements that exist today.

Most operators, on making their first call, will identify the station they are calling, although this is not required by the rules. They may also identify other stations in the group, if this seems appropriate, then drop most of the identification of call signs except their own after that, until on their final transmission. This procedure seems to work the smoothest, but it is a matter for each operator to work out for himself, since the FCC rules leave the subject wide open.

Some busy nets establish ID procedures for conducting net operations, and this is administered in various ways, depending on the sternness of the net manager. In any event, there is no need to identify every minute or half minute during the rapid exchange of transmissions that is characteristic of ssb and f-m operation today. Operators recognize each other by voice, so a simple "Go ahead, Joe," or "Hey, Bill" usually will suffice in informal conversations, and will be perfectly legal as long as each station identifies *itself* each 10 minutes, and then identifies *itself and one other station* at the end of its operation.

MOBILE AND PORTABLE IDENTIFICATION

At the same time that it changed the requirements for general station identification to the newer abbreviated system, the FCC made sweeping changes in mobile and portable ID rules. For many years, it was possible to stir up quite an argument on the air when you heard a station signing as "mobile six" or "portable eight" or the like. Formerly, there was no such designation, but many operators apparently thought there was because they would see calls listed in magazines such as W7DQS/9 or W1NJM/3. This is the form used then and still used on cw, and it is the simplest and most practical way to list such stations on paper. But on the air on phone, there was a different requirement—it was that a station should state "portable" or "mobile" after his call sign, and then give *the approximate geographical location* of operation.

There never was any clearcut explanation of how much precision was needed in giving this location, so everybody pretty much made his own rules in this regard, depending on his circumstances and his conscience. So, the correct ID under the old system would be like this: "This is W7DQS portable at the Vulture Mine near Wickenburg, Arizona," or "This is W1NJM mobile on the Pennsylvania Turnpike 20 miles north of Allentown" or something to that effect.

On cw or rtty, the geographical location was not required, only a slant bar and the number of the call sign area in which the operation occurred. This discrepancy between phone and cw rules bothered hams for years, and caused a lot of confusion and more than a few arguments on the air regarding what was proper. Finally the FCC settled the squabble by throwing out the old rules and writing new ones that pretty much conformed to what most hams had been doing anyway.

Describing the "old way," which is no longer applicable, may seem like wasted time now, but most operators today can benefit from understanding the old method because they will hear various versions of it still in use on the air.

The "new way" is for portable and mobile stations to identify as "portable six" or "mobile eight" or whatever the situation may be, and for cw and rtty stations to continue using the slant bar and local call area number just as they always did. It is no longer necessary for mobile or portable stations to give their location, other than the number of the call sign area in which they are operating at the moment.

This method certainly makes mobile operation much smoother, and eliminates a lot of excess talking that used to be required. It also brought phone and cw rules more nearly in line, since under the old rules you could operate cw (or rtty) while mobile or portable without giving your geographical location, yet you were required to give this location when on phone.

Another welcome change in mobile rules which now is pretty well understood is that it is no longer necessary to enter in the log the exact time you began and ceased working each individual station with which you talked while mobile. Serious operators used to keep a clip board beside them on the car seat and try to jot down the time of each contact when made on the air. This could be quite hazardous in traffic, and so was wisely dropped.

Although a mobile operator still needs to enter the call sign of each station he works, the only times that need be entered in the log are the starting and ending times of mobile operation. This relieves the pressure of attempting to write down a series of times while driving. It is usually not hard to recall call signs you have worked during any given stint behind the wheel and enter them in the log when you reach a convenient stopping place.

An inconsistency between FCC logging and ID requirements rears its ugly head when a ham operates mobile, though—you do not need to give your location on the air as you identify, but you *do need to enter it in the log*. There is nothing in the rules to indicate how much detail is required in describing the location, however.

This same situation applies to portable operation. Giving the location is not required on the air, but it is required in the log. And in portable operation, just as in operating at the home station, it is required that you enter the starting and ending times for each station you worked, rather than just the overall time period as may be done in mobile logging.

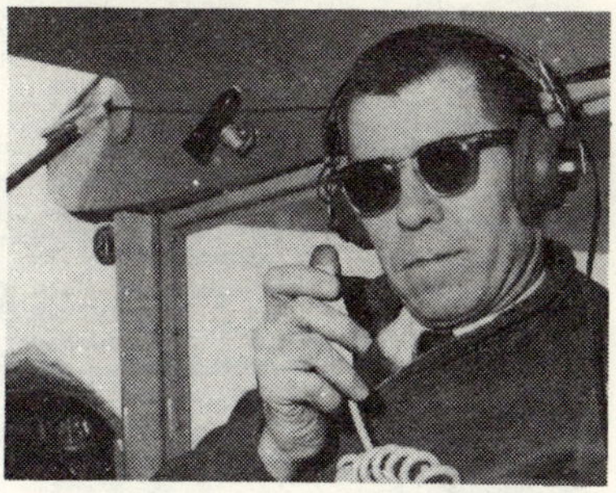

Amateur radio operation on board light aircraft is becoming increasingly common. Low-power transmitters can be heard great distances when airborne. Airborne operation is considered by FCC rules to be the same as operation in an automobile.

Incidentally, the question is often raised, just exactly what is considered "mobile" and what is "portable?" And what about those exotic-sounding identifications you hear of "maritime mobile" and "aeronautical mobile?" "Portable" operation has been considered any operation at a location which is different from the location shown on the station license, but which is not operated in a vehicle.

Some fellows who move temporarily, maybe while attending college or while working temporarily at a job in a new location, will identify as being "fixed portable." There is nothing in the rules to substantiate such a designation—either a station is "portable" or it is not. There are no degrees of being portable.

"Mobile" operation is considered operation of a station in a vehicle so that the station is capable of being operated while the vehicle is in motion. This vehicle may be a car, truck, bus, sand buggy, motor boat, airplane, steam locomotive—anything that rolls, flies, slides, or swims qualifies. Even if you are sitting in your own car parked in your own driveway, the station is considered a mobile station if it is capable of being operated while the vehicle is in motion. So, the occasionally heard "mobile at rest" and "mobile in motion" have no basis in the rules; they just seem to be the product of some fellows' overactive desire for colorful descriptions of their operation.

The terms "aeronautical mobile" and "maritime mobile" sometimes are heard from stations operating in an airplane or on board a boat, but again these terms have no basis in the FCC rules. (Formerly, these terms were required on international flights or voyages only.)

Portable operation shown here being used to assist in a charity drive, is any operation away from the location shown on the station license, except for operation in a vehicle, which is termed mobile operation. Rules for logging and identification of these three types of operation differ in details.

Incidentally, when operating portable or mobile more than 48 hours away from the home-station location shown on the license, you are supposed to notify the FCC office for the district in which you live, and for each FCC district in which you operate, giving your location, and in the case of mobile operation, a description and license number of the vehicle in which the equipment is used.

However, if you keep your trips under 48 hours, you can dispense with this paperwork, since no notification of any kind is required by the regulations.

RTTY ID REQUIREMENTS

A few special requirements exist for identification of amateur rtty stations. Rtty is similar to cw in that it utilizes pulses of carrier signals to convey information, but the pulses do not correspond to Morse Code. It is impossible to listen to an rtty signal with the ear and determine what is being sent; only a teleprinter and its associated electronics can "decode" the signal.

For many years, the FCC has required hams operating rtty to identify their stations both by rtty and cw. The reason given has been that FCC's mobile monitoring stations are not equipped with teleprinter machines, so the only way a mobile monitor could determine the identity of an amateur rtty station is by hearing that station's call letters on cw.

Originally, FCC required that the rtty station send on cw both its own call sign and the call sign of the station to which it was sending. Later, the rule was changed so the transmitting station needed to send only its own call letters on cw, but could send the call sign of the station with which it was communicating on rtty. The advantage of this, in modern, efficient rtty operation, is that a station operator can set up a mechanical means of sending his own call sign on cw automatically at the push of a button. However, it is not practical to try to do this with the call sign of every station he works on the air.

When the FCC changed its rules to provide for the newer, greatly abbreviated method of identification of cw and voice stations, it also deleted the requirement for rtty identification of the station to which a transmission is being sent. Now, the cw identification of its own call sign is the only ID required of rtty stations, even at the end of a series of transmissions.

F-M REPEATER ID

An additional special case of ID procedures occurs with a specialized type of f-m operation. Many hams who have taken up f-m operation have built remote stations, at elevated locations such as mountain tops and atop tall buildings to extend their operating range on the 2- and 6-meter and 450-MHz bands, where the f-m operation exists.

Two types of these stations are in use. One, called a "remote base," is simply a transmitter and receiver which can be operated by remote control (usually by a 450-MHz radio link) from the operator's home station or mobile station. The other type is called a "repeater." It also consists of a transmitter and receiver, which picks up a signal from any home or mobile station on a specific frequency and retransmits it on another frequency, usually in the same band.

In both cases, the remote transmitter has its own station license and call letters. Identifying these call letters on the air generally has been done in one of two ways by f-m ham operators. One way is to have an automatic code wheel, or solid-state code-character generator, which is turned on automatically by a timer or other means to transmit the station call sign in Morse Code when required.

Another way is for hams using either a remote base or a repeater station to give the call sign of the remote installation orally as they identify their own station. An example would be: "WA7KUM THIS IS W7KWB THROUGH WA7HUH." For special ID requirements outside the United States, *see* Chapter 10.

LOGGING REQUIREMENTS

Each amateur station must maintain a log, which must be kept on hand for possible FCC inspection for one year after the date of the last entry. Certain items of information are required in these logs, and additional information of interest may be included at his option.

The required information is: date and time of each transmission, signature of operator, call sign of stations called, input power to final transmitter stage, amateur band used, type of emission, location of station, and message traffic handled. (Stations which handle formal message traffic must keep these messages on file for one year.)

It is not necessary to put down the operating frequency—all that is required is the *band* in which operation took place, although the specific frequency within that band may be used if the operator wishes. When operating portable or mobile, the location at which the operation occurred also is required by the FCC rules. And mobile logs also are supposed to show the identification of the vehicle in which operation occurred.

When certain information is repeated over and over, it is not necessary to enter it for each transmission. It may be entered in the front of the log once to serve for the entire log. For example, many hams operate mobile on only one band, so they may enter this in the front of the log and not bother repeating it for each transmission. If a switch is made to another band, this may be entered on the appropriate page when the switch is made.

A peculiar inconsistency exists in the rules regarding situations in which someone else is allowed to talk over your station. Another ham may operate your station, with your permission, and must sign the log when doing so, indicating the transmissions he made. A non-ham may talk over your phone rig, or type on the keyboard of your rtty station, as long as you control operation of the transmitter and put the person's name in the log. But, a non-ham is *not allowed* to operate the telegraph key of your cw station, *even under your supervision and control.*

Incidentally, when another ham is operating your station, both you and he must be licensed for the type of operation conducted. For example, if you have a General Class license and he has a Novice license, he is allowed only to operate 75 watts crystal-controlled within the Novice band segments when using your rig. If your visitor has an Extra Class license and you have a General Class license, he must operate within the General Class band segments while using your rig. In the case of radio club members using a club station, both the individual members and the club station trustee must be licensed for the particular type of operation conducted, just the same as outlined in the preceding examples.

TAPE RECORDER LOGS

Some hams use a tape recorder to do their log keeping for them. This is especially handy on f-m because it is fairly simple to turn on a tape recorder with a relay keyed by the f-m receiver's squelch circuit so the recorder will come on whenever a signal appears at the receiver. By dictating the date onto the tape at the start of each day's operation and by dictating the time at the beginning and end of transmissions, or through use of a coded signal from a timer or WWV receiver fed to the tape, it is easy to comply with the day-to-day logging requirements. Other information such as band and power used and mode of operation may be added to the tape, which, of course, must be kept on file for a year after the date of the final transmission logged.

Chapter 3

Signal Monitoring and Reporting

Evaluating and sending reports of signals of the stations you work on the air is one of several aspects of radio *operating*, as distinguished from mere radio *using*. To do this job competently requires a moderate amount of skill in using a receiver, a working understanding of the operation of a transmitter, and a basic understanding of the effects of signal radiation and propagation. This, after all, is one of the skills which experienced hams are expected to possess.

This ability separates the men from the boys, and is among the factors which set experienced hams apart from users of the many other radio services which exist today. Most other radio-system users today are not required to have any technical knowledge, and virtually no operating skill.

To some nontechnical radio-system users, there are only two kinds of signal reports: either 10-2 (meaning "receiving you well"), or 10-1 (meaning "receiving you poorly"). For them, there just isn't any middle ground, in types of signal conditions and ability to understand transmissions. Users of these radio systems generally are not knowledgeable enough regarding operation of their equipment to give any helpful information to another user who is experiencing trouble.

Experienced hams, though, and I stress the word "experienced," by having an understanding of the technical workings of their equipment and of the atmospheric and geographical effects on signal propagation, can give signal reports which convey much more information. Acquiring this ability is part of the continuous process of learning which all serious ham operators accept as a part of their activity.

PURPOSE OF SIGNAL REPORTS

To be able to give worthwhile signal reports, it is best to understand why these reports are given in the first place. One thing which signal reports definitely are *not* is an exercise in "compliment club" tactics. Sometimes you will hear a signal report given as if out of a desire to inspire good will and rosy friendship in the operator to whom the report is given. That operator may respond with, "Thanks for the fine report, old man," which just further increases the misunderstanding of the real purpose of a signal report.

The primary function of a signal report in practical radio communications is to enable the sending operator to understand how well he is being received by the receiving operator, so that he can adjust his sending accordingly. If an operator is told he is being received well, he realizes his signal is not being clobbered by noise or interference, or subject to severe fading. Consequently, he has little doubt that he will be understood properly by the receiving operator. However, if he gets a report that his signal is weak, or that noise or interference or fading is causing the receiving operator to have trouble hearing him, then he knows he should repeat important parts of his transmissions and use phonetics if he is operating phone, or possibly spell out some words he might otherwise abbreviate if he is on cw.

A second important purpose of meaningful signal reports is to enable an operator to learn if he has any signal defects which he may not already be aware of. Such troubles as audio distortion in a phone transmitter, chirp or key clicks in a cw rig, or drifting of the carrier frequency in either type of rig should be included in a signal report, for the sending operator may not otherwise be aware of them.

Builders of home-brew gear, such as these radio club members admiring a home-made 6-meter f-m transceiver, rely on accurate, meaningful signal reports from other serious operators to verify results of their handiwork. Signal reports, to be useful, must give full information on both good and bad characteristics of signals.

Some hams seem hesitant to report these signal defects, for fear of offending the operator of the station with the trouble. There should be no such fear, for personalities should not be permitted to enter into the spirit of giving or receiving signal reports. A serious operator makes his reports honest, straightforward, and factual, and gratefully receives reports of any trouble he has in his rig.

Two additional purposes served by signal reports are help in determining how well a certain piece of equipment is working, and determining the effect of propagation conditions existing at the time of the transmissions. An operator may modify the audio system in his ssb transmitter, or the keying circuit in his cw rig, for example, and seek signal reports to determine how his signal quality has been affected by the changes he has made. Another operator may put up a new antenna, or readjust his directional array, and then seek reports to learn what this change has made on the strength of his signal.

In the case of the vagaries of ionospheric conditions which affect the transmission of radio signals, signal reports are the only means a ham has to learn how these changes in conditions are affecting his own transmissions. With honest signal reports from different locations, he can judge the type and extent of the propagation he is experiencing. For example, these reports are a great help to the ham who is studying sporadic-E, aurora, or meteor-shower propagation effects to learn how signals from his location are affected by these phenomena. The ordinary rag chewer or contest operator can pursue his activity more enjoyably if he knows, through honest signal reports, how well he is working into a certain part of the country, or of the world. Day-to-day variations in propagation conditions are affected in predictable ways by the effect of solar activity on the ionosphere (see Chapter 6); the current status of these effects is revealed in signal reports obtained from different locations at various times of the day over a period of a few weeks. The vhf-uhf experimenter learns much about the effective range of his equipment from signal reports.

So you see, honest and comprehensive signal reports are important to all phases of successful radio communications and experimenting, and to be useful they must be competently given and carefully interpreted when received. You might make a fellow feel good by telling him he has a strong, clear signal, even if it is actually weak and distorted, but you will actually be doing him more good if you told him honestly just how his signal sounds, so that he can make any corrections that are called for in his equipment, or understand how successfully he is able to work into the part of the country where you are located. A weak signal may be merely caused by the propagation conditions existing at the moment, especially if other signals from that area also are weak. If other signals are strong, however, a single weak signal could be indicative of trouble in the transmitter or antenna. A distorted signal almost certainly indicates trouble in the transmitter, which should be corrected as soon as possible. An amateur might not be aware of this problem unless told about it in an honest signal report from a fellow operator.

MEASURING AND REPORTING SIGNALS

A standard system of signal reports used by hams has evolved down through the years, providing the basis for exchanging signal reports which are universally understood. This system was originated for use on cw, but is readily adaptable to voice and teleprinter operation as well.

It is called the "RST" system, with the letters standing for readability, strength, and tone. When used in voice work, the tone portion of the report is dropped, with only readability and signal strength given. ("Tone" used in the RST sense does not refer to bass and treble characteristics of the audio modulation, but rather to the purity of the audio note produced in a receiver by a received carrier beating against the receiver's local oscillator, as in cw and rtty reception.) In rtty operation, the full RST is used, and sometimes an estimate of "percentage of copy" or "percentage printed" is added.

A series of digits are used to make up the full RST report, with each digit having a specific value for readability, strength, and tone. These standard values are:

Readability

1—Unreadable.
2—Barely readable, occasional words distinguishable.
3—Readable with considerable difficulty.
4—Readable with practically no difficulty.
5—Perfectly readable.

Signal Strength

1—Faint; signals barely perceptible.
2—Very weak signals.
3—Weak signals.
4—Fair signals.
5—Fairly good signals.
6—Good signals.

7—Moderately strong signals.
8—Strong signals.
9—Extremely strong signals.

Tone

1—Extremely rough hissing note.
2—Very rough ac note, no trace of musicality.
3—Rough, low-pitched ac note, slightly musical.
4—Rather rough ac note, moderately musical.
5—Musically modulated note.
6—Modulated note, slight trace of whistle.
7—Near dc note, smooth ripple.
8—Good dc note, just a trace of ripple.
9—Purest dc note.

As you can see from the table, the best possible signal report you can give is "599," which would mean "perfectly readable, extremely strong signals, with purest dc note." Although you often hear 599 reports being given by overly enthusiastic operators on making contact with a station that is fairly strong and easy to copy, there actually are few signals that merit a 599 report. A typically good, clear signal usually is a 579 or 589. If you receive that kind of report from a serious operator who makes careful, accurate signal reports, you can be assured you have a good signal at his location and he is copying you with ease.

Some operators tend to overlook the fact that readability of "5" means perfect copy—the same as "R" on cw or "roger" on phone. (see Chapter 1). So, when the report you receive includes a readability of "5," you know that the operator who gave it understood your entire previous transmission. Keep this in mind when giving signal reports too, so you can make your reports accurate. And don't be hesitant to give a readability "4" report if you didn't hear the other fellow's transmission well enough to copy everything. Likewise, don't be hurt if you are given a readability "4" report. You will see, when you study the preceding RST table, that a report of "479," or even "469," really isn't bad. A "469" report means the receiving operator copied almost everything you sent—maybe missing only a few letters—and that your signal is not exceptionally strong, but still is considered good.

Assuming your equipment is working properly, there is usually nothing you can do about it if your signal is reported as being weak. You probably are just an innocent victim of propagation conditions. Incidentally, keep in mind that "readability" in a signal report does *not* refer to any flaws in the other operator's cw fist, or his personal voice characteristics on phone. These you will have to deal with as tactfully as you see fit.

As far as your own equipment is concerned, generally the only time to get concerned is when you receive a report that your tone is less than "9." This generally indicates that something has gone wrong in your transmitter power supply, causing some ac ripple to appear on the B+ voltage. A tone of "9" means you have a pure note—a "smooth" carrier with no trace of ripple. However, if a filter capacitor begins to fail, or a filter choke is not working satisfactorily, the dc delivered to the transmitter's oscillator, driver, or final stages will no longer be pure dc, but will have a little ac superimposed on it. This causes a rough-sounding note on cw. The amount of roughness is indicative of the amount of ripple present on the carrier, and this is what is covered by the tone report. Any report that his tone is less than "9" should alert an operator to check his transmitter power supply immediately, for hum on the carrier is annoying to receiving operators, as well as being contrary to FCC regulations.

Be careful in giving reports to another station, however, because occasionally an interfering signal or some radio noise will mix in your receiver with a weak signal and cause the weak signal to appear to contain ripple, when actually it does not. Whenever you are not sure about the source of apparent ripple you notice on a signal, tune back and forth across it to see if the character of the hum changes. If it does, this is an indication that it is being produced by interference, rather than by ripple on the dc power in the transmitter.

It has become customary to give RST reports on cw or rtty this way: "UR RST 579 IN . . ." with your location following the word "in."

On phone, the words "solid copy" generally are used to indicate a readability of "5." Of course, as already mentioned in Chapter 1, the word "roger" also carries this meaning. Signal strength reports on phone generally are given on the basis of the indication of the "S meter" on the receiver, rather than on the basis of estimates made by ear. More will be said about that unfortunate source of errors and controversy just a bit later in this chapter.

Some phone operators give a solid-copy readability report as being "Q FIVE," and some give a report of solid copy and good signal strength as being "FIVE BY FIVE," or sometimes "FIVE SQUARE." These are expressions picked up in military service, or from imitations of other operators, and do not conform to standard amateur practice.

Along with the RST report on cw or rtty signals should be two other items, *if* certain conditions exist. These are *key clicks* and *chirp*. Key clicks are caused by a characteristic of the transmitter oscillator whereby the oscillator creates a sharp pulse as it is keyed. This causes a clicking noise in the receiver, which can prevail on frequencies above and below the one you are operating on, sometimes even outside the amateur bands, causing interference to other stations, both amateur and commercial. If an operator detects key clicks on a signal he is receiving, he adds the letter "K" after the digits of the RST report he gives. Such a report might be "RST 579K."

A chirp, on the other hand, is caused by an unstable oscillator. This may be a fault of the oscillator stage itself, or it may be caused by poor power-supply regulation in the transmitter.

Whatever the cause, the result is a cw note that shifts slightly in frequency during the time the key is closed, causing a "chowpy-chowpit" sound instead of a pure tone. When hearing such a signal, the way to report it is to add "C" after the RST digits, such as "579C."

Either of these reports indicates a fault in the transmitter of the station to which they are sent, so this type of report is intended to tell an operator of a defect in his signal which he may not be aware of—at least not yet. Both these items also are violations of FCC regulations, because they can cause interference to other stations, so they indicate the need for immediate repairs.

Once in a great while, you may hear a signal report with "X" following the RST numerals. This indicates the signal has the stability characteristic of crystal control. This report is seldom used nowadays, since most VFOs are capable of this type of stability routinely.

Q SIGNALS FOR SIGNAL REPORTING

Useful along with the RST signal reporting system are some Q signals which can give an operator whom you are working a more complete idea of receiving conditions at your location, and of the characteristics of his signal.

"QRM" is a signal that has become so familiar on the crowded ham bands that it is often erroneously referred to as a "condition." That is, a ham often will say, "The QRM tonight sure is bad." This is an obvious misuse of a Q signal, and yet it is one which has become universally understood, so it appears we will all have to learn to live with it.

All Q signals are complete statements (or questions) in themselves, with no added words needed to complete the meaning. "QRM," as shown in older lists of Q signals, meant "I am being interfered with," and "QRM?" meant "Are you being interfered with?" As used today on the ham bands, however, it means "Am I being interfered with?" when asked as a question, and "You are being interfered with" when used as a statement. This does not necessarily refer to *deliberate* interference, but to any interference which makes receiving the desired signal well difficult.

"QRN" is often used during periods of storm activity. It means "I am being troubled by static." When used with a question mark (QRN?), it means "Are you being troubled by static?"

These two Q signals often are used along with an RST signal report because they give the other operator an idea of receiving conditions so he can adjust his sending to make the best of the difficult situation.

Other Q signals useful in connection with signal reports are:

QRG—Your frequency is . . . kHz.
QRH—Your frequency varies.
QRI —The tone of your transmission is . . . (1=good; 2=variable; 3=bad). Note that in this case, a numbering system is used which is different from the 9-point table of tone values used in the RST system.
QRK—The legibility of your signal is . . . (1 to 5).
QSA—The strength of your signal is . . . (1 to 5).
QSB—Your signal is fading.
QSZ —Send each word or group twice (generally used in traffic handling during poor radio conditions, also useful occasionally in ordinary conversations).

S-METER USAGE

Most communications receivers are equipped with "S" meters, and hams have grown accustomed to using these meters as sources of signal-strength reports in voice work. Unfortunately, careless use of these meters can produce signal reports which are of little value.

These meters generally are connected to the receiver's avc circuit in such a way that stronger signals produce higher meter-scale indications and vice versa. The meters usually are calibrated in S units from 1 to 9, and in dB above S9. It has become an unfortunate habit of some voice operators to take these meter indications at face value, without pausing to understand just what the meter is telling them.

Actually, the only thing an S meter is capable of doing is indicating *relative signal strength.* In other words, the S meter can be a useful tool to use in comparing two or more signals, but it is incapable of measuring the *absolute strength of one signal,* unless the meter and all receiver controls which affect the meter's indications have been calibrated against a signal of known strength.

A second characteristic of S meters which is often overlooked is the fact that the receiver's rf gain control has an effect on the meter indication. Prove this to yourself by tuning in a signal with the rf gain control fully "on," and note the S-meter indication. Then reduce the rf gain control setting, and you will notice the S-meter indication also is reduced, yet the incoming signal is the same strength. So, you can see how meaningless it is to tune the receiver to a signal and report that the station is "30 dB over S9," for instance.

However, you can put the S meter to good use in *comparing* two or more signals. Let us say you are working two hams who are both in the same city, and you want to report which of them has the stronger signal, and how much stronger it is than the other. In this case, the meter will give you a valid comparison.

You might find one signal is S5 and the other is S8, so you can truthfully report that the stronger signal is three S units stronger than the weaker. The same type of comparison can be made on the dB portion of the S-meter scale.

There are two precautions which you must be careful to observe when you do this:

First, be sure you have each station properly tuned in before noting the S-meter indication, because the two stations may not be precisely on the same frequency, so your receiver might favor one over the other until it is retuned.

Second, do not change the rf gain control setting, or the i-f selectivity control setting on your receiver while you are comparing signals. To do so would change the signal within your receiver being measured by your S meter, even though the incoming signal at your antenna did not change.

An S meter can also be very useful in several ways as a sort of combination tuning indicator and field-strength indicator. As you tune voice stations, for instance, you can observe the meter to determine when you have a station tuned in for maximum signal. You also can use it when rotating your beam antenna, to determine when you have the antenna pointed properly to receive a desired signal with maximum strength. You can also use this method to determine the approximate direction from your location of a station being received. An S meter also can be useful to measure relative signal strength when you are doing alignment work on your receiver.

MONITORING YOUR OWN SIGNAL

Checking the operation of your transmitter by listening to it with your receiver is an operating convenience often overlooked. This procedure has several advantages, and once an operator has tried it, he is seldom satisfied to go back to the older method of transmitting without any way of knowing at his own station how his signal sounds.

Of course the receiver cannot be directly connected to the antenna at the same time as the transmitter when transmitting, but the receiver does not necessarily need to be muted when you transmit either. In voice operation, it is generally the practice to mute the receiver with a relay or solid-state switch whenever the transmitter is put on the air, so as to eliminate the feedback that would occur if the receiver speaker was operating while the mike is "live." Often the switch that is used for muting is actually a second set of contacts on the antenna changeover relay. However, the innovative ham often will wish to bypass this muting system, at least part of the time, so he can listen to his own transmitted signal to check its quality. On phone, this may require using headphones on the receiver output in place of the speaker during such monitoring to prevent audio feedback from speaker to mike. Then, all you need to do is disable the circuit which normally mutes the receiver during transmissions.

Using a tape recorder connected to the receiver output will give you a record of your transmitter performance, and allow you to compare the effects of circuit changes and of different microphones. You may need to try more than one recorder to get one that will work in this application, because some tape recorders are adversely affected by the

Many amateurs operate both a home station and a mobile station. Competent signal reports from other amateurs render valuable assistance in both types of operation by assessing effects of equipment modification and adjustment, and by giving first warning of trouble in the rig or antenna system.

presence of the rf field created by a nearby transmitter. In some cases, the tape recorder audio becomes distorted, while in others, the recorder motor-speed control circuit is affected, so that the tape recorder motor races when the transmitter is turned on.

A method commonly used by cw operators, especially by traffic handlers, and also by some rtty operators is to use an electronic TR switch (transmit-receive switch) between the transmitter and the receiver. The transmitter is connected to the antenna all the time, while the TR switch isolates the receiver input from the transmitter during transmit periods to prevent damage to the receiver front end. No receiver muting is used in this case, since there can be no audio feedback in cw or rtty operation, since there is no microphone used.

With this simple and very useful system, a cw operator not only listens to his own signal all the time, but has the added advantage of being able to hear any changes in band conditions or any interrupting stations on the frequency whenever his key is open. This provides full "break-in" capability, which can be very useful in cw operating.

Many of the ssb transceivers, which have become so popular on the ham bands in recent years, also will work on cw, but have only limited break-in capability. Generally, the VOX relay controls the switching of the unit from receive to transmit and back to receive. When the key is closed, the VOX relay closes and puts the rig on the air. However, the relay-control circuit generally is adjusted to hold the relay closed for a second or two after the key is opened. Consequently, the operator cannot hear another station trying to break in unless he pauses long enough for the VOX relay to open and switch the unit back to receive. This is better than the older way of using an outboard antenna-changeover relay and not switching the relay to receive until the end of a complete transmission—however, it is still a poor substitute for full break-in capability if you work a lot of cw and want the full advantages that are possible.

Monitoring your own signal in your receiver as you transmit has another advantage that from a practical standpoint is impossible to achieve any other way; that is, the capability of being able to sense any drift in your own transmitted signal. You also will hear any chirp that might occur in your rig.

On rtty this monitoring ability allows the operator to print out his own transmitted signal on a teleprinter machine in his shack, so he has a constant check on the intelligibility of his own signal.

Some modern transceivers contain a "sidetone monitor" which provides an audio tone whenever the transmitter is keyed. It is important to realize this tone is *not* a reproduction of the transmitted signal, so it cannot be used to check the quality of the transmitted signal. It is useful as a monitor of the operator's "fist," but that is all.

OSCILLOSCOPE MONITORING

An oscilloscope may be used to monitor your own signal while you transmit, and also may be used to monitor signals you are receiving, so you can give more accurate and meaningful reports to stations you work.

Several methods may be used, with none of them requiring anything more than a simple, economy-priced scope. For monitoring of transmitted signal *only,* of cw, rtty, or ssb transmissions, the scope vertical input is connected through a suitable blocking capacitor to the antenna feedline, in parallel with the antenna. By using internal sweep to produce a scope trace, you will have a graphic picture of the rf envelope emitted by your transmitter.

With an a-m transmitter, connection also is needed to the modulator in the transmitter, so you obtain a scope pattern that accurately indicates your modulation percentage. For precise details on how to make these connections and how to interpret the scope display, check your transmitter and oscilloscope manuals, or refer to such reference books as the *ARRL Handbook* or the Editors and Engineers *Radio Handbook.*

An alternate way to connect the scope, so you can monitor *both* transmitted and received signals, is to connect the scope vertical input to an i-f stage in your receiver. Use a small value capacitor in series to prevent degrading the performance of the receiver i-f amplifier. Again, internal sweep is used to produce a trace on the cathode-ray tube face. This method will give you a trace which is representative of the signal envelope in the i-f portion of your receiver, so you can analyze all signals you receive on the air.

By listening to your own transmitted signal with your receiver, you can at the same time see your own signal on the scope, as well as other signals on the frequency to which the receiver is tuned. You will be able to recognize flattopping of ssb signals, key clicks, and ripple on cw and rtty signals, and even have an instantaneous and precise display to use in judging relative signal strengths. Since your own transmitted signal will be much stronger than those you are receiving you will need to reduce the rf gain to keep the scope display from running off the top and bottom edges of the CRT, but otherwise the monitoring of your own signal will be identical to monitoring of other signals on the band.

Another useful scope-type monitor used by some hams is a *spectrum analyzer*, or *spectrum monitor*. This is a special type of scope, built for this one purpose, that also is connected to the i-f section of your receiver. It displays signals across a selected portion of the band, and can help you find parasitic oscillations from your own or someone else's transmitter, and spot the location on the band of stations which are causing interference with excessively wide signals. (It also can be useful to spot an unoccupied spot on the band for sending a CQ, or for finding the heavy activity during a contest.)

GETTING HELP IN MONITORING

In some cases, monitoring of your own signal will not always tell you all you want to know. For example, it is hard for a ham to be absolutely certain from tests in his own shack that he is not radiating harmonics. Every transmitter has harmonics, but in a properly-designed and tuned rig, they are so weak they are not radiated far enough to cause other stations trouble. However, in your own shack, you may hear many harmonics of your signal in your receiver.

So, you need a helper—a competent and trusted ham friend who lives a number of miles away and has good receiving equipment covering the frequencies in which you are interested. Make contact with him by radio or telephone, tell him your operating frequency, and work out arithmetically several harmonics of that frequency. Then have him listen on those harmonic frequencies for your signal. If he hears none—and with modern equipment he normally should not—then you can operate with greater peace of mind. If he does hear some, he can work with you to verify when you have debugged your gear.

MONITORING AND MEASUREMENT TIPS

A familiar hassle among hams involves disagreements over who is "on frequency" and who is "off frequency."

Except for some DX work, where different transmit and listening frequencies are used deliberately, most ham operation calls for sending and receiving on the same frequency, and for all stations in contact with one another to transmit on the same frequency.

Unfortunately, disagreements (to put it mildly) occasionally erupt over the degree of precision with which this delicate maneuver is accomplished.

If you can monitor your own transmitted signal in your receiver, or at least monitor your own VFO output in the receiver before putting the rig on the air, you should be able to transmit precisely on the frequency of any station you can hear, *if* you can hear his carrier.

This is a big "if" in these days of widespread ssb operation, because a disadvantage of ssb (yes, there are a few) is that there is no transmitted carrier with which you can zero beat your own rig. At least there shouldn't be, since proper ssb operation calls for complete carrier suppression.

So, all you can go by is to tune your receiver to the most pleasing audio response of the voice you are listening to, then key up your VFO and adjust it to a zero beat with your receiver's local oscillator. If you want to get precisely on the same frequency of another station, you have to ask him to insert a little carrier momentarily in his signal to give you a signal to zero beat with.

With many modern ssb transceivers having both transmit and receive functions controlled by a common oscillator, you cannot even do this. You must tune in a station the best you can, and then transmit on whatever frequency that sets up your transmitter for. If the other operator also is using a transceiver with common transmitter-receiver oscillator control, and has a different idea of what your audio should sound like, he may adjust his receiver tuning (and in the process simultaneously change his transmitter tuning) so that when he makes his next transmission he will be on a different frequency. Then if you readjust your tuning, you will do the same thing. Get a couple guys doing this, and in the course of an hour or so they can slowly walk right across a phone band.

The same thing can happen on cw, when using a transceiver having common control of both transmit and receive frequencies. So when using this type of gear, it is necessary to always keep in mind that whenever you change the receiver tuning you are making a corresponding change in transmit frequency. Some of this type of gear now has available a second VFO so that transmitter and receiver tuning may be done separately when desired.

In zero-beating an a-m, cw, or rtty signal, it is necessary to remember to tune for a *full zero beat,* and *not* for a *beat note.* Some operators have the mistaken impression that whenever they hear the tone from their VFO in their receiver, that they have the transmitter tuned to the same frequency as the receiver. This is emphatically not so! To get the VFO on the same frequency as an incoming signal, you must adjust the VFO until it beats with the incoming signal and produces a dead zero beat—*an absolute audio null.* If you switch off your BFO and use the VFO to beat against the incoming signal frequency, you will hear a high pitched note in the receiver. As you tune closer, this note will become lower in tone. Keep going until the note drops so low in tone that you can't hear it—then you have achieved a true zero beat, at least within the limits of human hearing.

If the signal from your VFO is so strong that it drowns out the incoming signal in the receiver, then you can use a different approach. First, without keying on the VFO, adjust the BFO control on your receiver so the beat produced with the incoming signal reaches a null. Then, *without changing the receiver tuning,* key on the VFO and adjust it for a null in the receiver. Now you have the transmitter on the same frequency as the incoming signal. Return the BFO control to its original setting, and you will find the incoming signal and your transmitter produce the same audio note in your receiver—further proof that they are on the same frequency.

Either of these procedures may be used when using a precision external oscillator for frequency measurement. The precision with which you do this will have a lot to do with your ability to make accurate frequency measurements of signals you hear on the air.

This method also gives you the opportunity to determine if signals you hear are fundamentals, or are "images." (An image is a "false signal" produced in your receiver through mixing of a signal outside the receiver passband with a local oscillator to produce a heterodyne that falls within the receiver passband.) Sometimes you may hear a station, either ham or commercial, appearing to operate on a frequency not permitted for that particular operation, then later find you were hearing an image instead of a fundamental.

With a variable external oscillator, such as a tunable frequency meter or a VFO, as long as it is not too strong, you can quickly separate the images from the fundamentals. You do this by tuning the external oscillator so as to produce a beat *note* (not a null in this case) with the signal in your receiver, and then rock the receiver tuning back and forth across the audio tone. If the tone changes in pitch, it is an image; if it does *not* change, it is a fundamental.

This method is often used by ARRL Official Observers to detect stations operating in certain band segments for which they are not licensed, and by members of the ARRL Intruder Watch to determine if commercial stations are operating illegally in ham bands. There is more information on these two programs in Chapter 4.

Chapter 4

Broadening Your Operating Horizons

When an amateur first gets on the air, most operating is done for the sheer novelty of being able to talk to someone many miles away by radio. However, after this initial thrill wears off, many operators begin to want to accomplish something with their radio operating, to polish their skills and proficiency at some specialty, or at least to improve their general operating procedures so they become "one of the boys." This chapter will summarize a number of operating activities and characteristics which can be worthwhile goals for serious operators. There is no better place to start than with the subject of good manners.

OPERATING COURTESY

There is nothing in the laws and regulations that govern ham radio requiring any operator to exercise courtesy on the air, just as there are no laws that prohibit you from scattering cigar ashes on the carpet when you visit in a friend's home.

But courtesy and thoughtfulness of others are just as important in radio contacts as they are in social contacts. Radio frequencies, especially those used by hams, are great big nationwide, or worldwide, party lines. Anything done on the air, even though it is originated in the privacy of your own home, immediately becomes a public display; and it can become a public irritant if done without thoughtfulness of others.

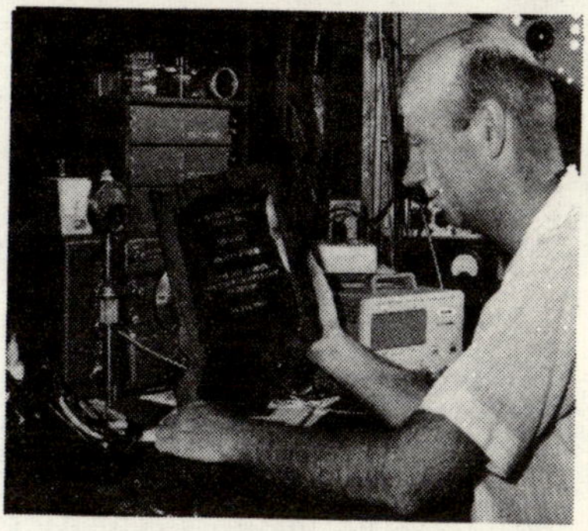

Many types of awards are given by national and local amateur radio organizations to operators who have proven themselves outstanding in various ways. Operating courtesy, participation in radio club activities, and helpfulness in public service projects are among work for which such awards often are given.

For this reason, and so that all amateur radio operators may have an equal opportunity to benefit from use of our frequencies, a number of "dos" and "don'ts" concerning operating courtesy have evolved. These rules of behavior have no legal power behind them, but nevertheless they are observed by serious amateurs just as if there was a severe penalty for disobedience.

Listen first is the number one rule which actually applies to all radio communications, not just to amateur communications. Hopefully none of us would barge into a roomful of people and interrupt whoever was talking. Yet this is what a radio operator does, in effect, if he begins transmitting without listening first to determine if the frequency is clear.

By listening first, before going on the air, you can determine if someone else, possibly with a weak signal, is already using the frequency, and so avoid causing interference to another station. If someone else is using the frequency, the courteous thing to do is wait for him to finish, or else move to another frequency that is unoccupied.

After you do go on the air, it is still a mark of courtesy to keep your transmissions reasonably short to give operators whom you are working frequent opportunity to reply to what you have said. Also, frequent pauses permit others to break in, if they wish.

If you have ever listened to a long-winded monologue from an operator who droned on for five or ten minutes at a stretch, you know how annoying a long transmission can be to other operators. So, reason demands that you show them the same courtesy that you would like them to show you.

So, the second rule of good operating courtesy is *keep all calls and transmissions short.* In making a CQ call, or a call to another operator who may not be familiar with you, courtesy, along with that most uncommon quality known as common sense, dictates that you not only do not make the call excessively long, but also that you give your own call letters a reasonable number of times so anyone hearing you will have ample opportunity to copy them.

It is surprising, and a little sad, to hear an inexperienced operator call "CQ" seemingly a hundred times, and then sign his own call letters only once or twice. A station hearing the call will recognize the CQ portion immediately, but would like to have the call sign repeated at least a couple times in case a burst of noise or interference prevented copying it fully the first time it was given.

Often you will hear someone refer to a "five by three" call or something of that nature. A "five by three" would

Amateur station W1AW at ARRL headquarters, Newington, Conn., makes cw transmissions of multiple-speed code practice, plus transmissions on cw, ssb, and RTTY of news bulletins daily for benefit of any amateur or amateur-to-be. The station also transmits code-proficiency runs and frequency-measurement tests, and is available at announced times and frequencies for informal QSOs with any amateur on cw, phone, or RTTY.

be five CQ calls, followed by the call sign three times, like this: "CQ CQ CQ CQ CQ DE W7DQS W7DQS W7DQS." Repeating this group three times generally is more than sufficient to snag a contact if anyone on the frequency wants to visit with you. There is no hard and fast rule that calls must be "five by three": the objective is that the call be long enough to be understood clearly, but not so long that it becomes a waste of everyone's time and a nuisance to other operators.

After making a CQ call, or even after calling a known station with which you have a prearranged schedule, it is wise to pause a minute or two and tune on both sides of the frequency on which you transmitted, to hear any replies that may not be exactly on your frequency. Sometimes you will hear an operator give a long CQ, pause briefly, and then start calling CQ again, while at the same time a couple other operators are calling him slightly off his frequency. It seems that some CQ callers expect answers to be loud and clear, and to be precisely on their calling frequency, and they do not make an effort to look for weak replies which may be on nearby frequencies. This is a luxury unwarranted by the facts of life in amateur radio. (Incidentally, when a station is calling another specific station, it is considered bad manners to call the calling station unless you have something important to tell him. He obviously wants to talk to someone else, so give him a chance to do just that.)

Speaking of keeping transmissions short, this is a display of operating courtesy which is especially appreciated by mobile operators. You may be sitting comfortably at home with your feet on the table, and so feel no need to avoid long windedness. But the mobile operator to whom you are talking may have finally fought his way through annoying traffic and reached his destination, where he needs to get out of his car for a short time, perhaps to go into a store, before continuing his drive.

If he has to sit several minutes in his car, draining his battery or overheating his engine, waiting for you to finish talking so he can wind up his contact with you, he is going to become understandably impatient with you. Also, if the

Hundreds of amateurs throughout the United States and Canada hold appointments in the field organization of the ARRL. A variety of appointments are available to match the wide variety of amateur operating interests. The objective is to encourage organized operation to provide improved public service and enable greater accomplishments by individual amateurs.

mobile operator happens to be from some other part of the country and he is unfamiliar with your area, he may want to ask you questions about his route as he drives along. If you don't give him frequent opportunities to do this, he may drive many miles out of his way, or get on a wrong road or into needless traffic congestion which you could have helped prevent if you had given him more chances to ask for information. It is also good to remember that a mobile operator has many distractions and interruptions, plus the overriding need to keep his mind primarily on traffic conditions rather than radio operating. The home-station operator has none of these things to contend with as he operates his station, so it is a mark of genuine courtesy for the home-station operator to keep these in mind while working a mobile station.

As you can see, radio operating courtesy, just as social etiquette, consists primarily of controlling your actions out of consideration for the other fellow. By observing courtesies such as described here, everyone can accomplish much more with his radio operating, and enjoy the experience more fully than if operating becomes a "dog-eat-dog" situation.

BREAK-IN CAPABILITY

When carrying on a face-to-face conversation, a group of people can interrupt one another (politely, of course) as appropriate in the course of the conversation. You may make a statement, then pause briefly for others in your group to voice their two cents worth. This is the way most conversations work, and a large amount of information may be exchanged rapidly in an ordinary fashion, as long as no one person monopolizes it.

This very same situation can exist in radio operating, if proper procedures are followed. Keeping transmissions brief is a key to this, for this allows other operators to transmit as they need to so a lively exchange of communication is carried on.

An important advantage is the capability of operating "break-in." This simply means that your transmissions are made in such a way that other stations may interrupt you if necessary, and this obviously requires that your station be set up so you can receive whenever you pause in your transmitting.

On ssb, this is easily accomplished by the use of VOX (voice operated transmit) which is built into most ssb rigs. When you begin talking into the mike, the audio circuitry in your rig trips the VOX relay which mutes the receiver and puts the transmitter on the air; when you pause in talking, the relay opens, killing the transmitter and restoring the receiver.

On cw, the same thing is accomplished by allowing the receiver to operate continuously, instead of muting it and disconnecting it from the antenna during transmitting periods. With this method, the transmitter remains connected to the antenna all the time, and the antenna is linked to the receiver with an electronic TR switch instead of an antenna relay.

When this is done, a cw operator can hear clearly anything on the frequency whenever his key is open. It is literally possible to "break" a transmission by an operator between the dits in the letter "H" if he is equipped this way!

This arrangement for a cw station has the added advantage of enabling the operator to monitor his own signal, as mentioned in Chapter 3. Also, it is good to keep in mind that the Q signal "QSK," in its latest-recognized meaning, refers specifically to a station equipped for cw break-in—that is, that the station is equipped so another operator may successfully break in on a transmission and be heard.

Many ssb transmitters and transceivers can be operated on cw with provisions for the VOX relay to disable the transmitter and restore the receiver whenever there is a pause in keying. This works the same as when operating phone, when the relay opens after a brief pause in speaking into the mike. However, the VOX relay circuit generally is adjusted to remain closed for a brief period (about a second or two) before opening. Otherwise, it chatters constantly, on either cw or phone, and becomes very annoying. So, with a VOX relay used on cw, the station is not truly equipped for break-in, because a station attempting to break would not be heard unless the operator of the VOX-equipped station paused a second or two in operating the key. It would be incorrect to say such a station is equipped for "QSK," since it is impossible to break such a station while the operator is sending.

Whether using true break-in capability with a TR switch, or using VOX on cw or ssb, the ability to *listen* during a pause in sending is a decided advantage to efficient and courteous operating. It allows an operator so equipped to hear at least some "breakers," as well as hear interference or noise building up on the band, or a change in propagation conditions occurring.

In addition to listening first, it is often wise to ask "Is the frequency in use?" before beginning a transmission, if in doubt. This might seem silly at first—you may think that if the frequency is in use, you would hear a signal, so if you don't hear a signal, then apparently the frequency is not in use.

But this is not always the case. On bands on which skip conditions prevail, you may hear only one end of a two-way exchange. Consider the example of a station 50 miles from you operating on 20 meters and working a station 1000 miles from you. You would hear the distant station, but you probably would not hear the nearby station, because of the propagation characteristics of 20 meters. At various times, this situation could occur on any amateur band.

So, if the nearby station was sending when you came on the frequency, you would not hear him and could assume the frequency was clear. However, the distant station could hear both you and the nearby station, so if you go blindly on the air, you might thoroughly clobber his QSO. All of us have been clobbered, at one time or another, by someone coming on to call "CQ" while we were trying to copy someone else. The first impulse is to get mad at the discourteous guy who busts on the frequency, but it could be that he *thought* the frequency was clear because he could not hear the station we were listening to.

This same type of situation can exist on vhf and uhf bands where f-m repeaters and remote base stations are becoming quite common. Often in these cases, it is possible to hear only one side of a conversation between a repeater or remote base station and a station not using remote facilities. When first coming on a frequency, it is wise to wait a while and see if there is any activity, or else make a brief call asking if the frequency is clear.

This situation with regard to f-m remote stations points up a special need for operating courtesy with this type of equipment. In case a repeater transmitter covers an area larger than the range of its companion receiver, it is easy for someone to start using the repeater and clobber other stations that are operating direct between each other. Because of the nature of f-m equipment, the strongest signal wipes out all others in a receiver. So, a strong repeater can totally wipe out weaker signals on the same frequency. If this frequency is one often used for direct station-to-station conversations, it has a very high interference potential.

RECOGNITION OF OPERATING SKILLS

Besides traffic handling, DX working and contest operating, each covered in separate chapters in this book, there are numerous other activities open to hams wishing to accomplish something with their radio operating abilities.

Often these accomplishments are related to improvement of operating abilities. For example, a "Frequency Measuring Test" intended to test individuals' skills at precision frequency measurement, is transmitted four times a year by the ARRL headquarters station, W1AW, on 80-, 40-, and 20-meter cw frequencies. Interested amateurs are invited to measure the frequency of these test transmissions, and the frequencies also are measured by a commercial measurement lab. Measurements submitted by individuals are compared with the commercial results. The measurement test provides a way for individuals to determine how accurate their equipment and techniques are, and also provides a standard for appointment of ARRL Official Observers, based on their demonstrated abilities (the Official Observer program is explained later in this chapter.)

Another test of operating skills is provided by the Code Proficiency transmissions made by W1AW, using machine-sent code at speeds up to 35 words a minute. Individuals who submit accurate copy made from these special transmissions each month are awarded a *Code Proficiency Certificate* indicating the highest speed they have copied.

Daily code-practice transmissions also are made by W1AW for the use of anyone studying for ham license examinations. These transmissions are made at speeds up to 35 wpm each morning and evening, on a schedule published in the ARRL magazine, *QST*. These transmissions have been used by many thousands of hams down through the years to improve their cw copying abilities in preparation for FCC exams.

This same station transmits news "bulletins" on cw, phone and rtty each evening on a wide range of items of interest to hams. The more important of these are repeated by numerous "Official Bulletin Stations," operated by volunteer hams in all parts of the country. By making a habit of listening to these transmissions regularly, a ham can be well-informed on events affecting his hobby in advance of publications of the monthly radio magazines. The W1AW cw bulletin transmissions are sent at 18 wpm, and so also make good code-practice material.

Besides the *Code Proficiency Certificates* already mentioned, other awards available as an indication of ability and activity include such certificates as *Worked All States* (WAS), *Worked All Continents* (WAC), DXCC (100 countries), and the newer five band WAS and DXCC awards, given for proof of contact with all states or with 100 countries on each of five bands.

Another highly coveted award acquired by a relatively few hams is membership in the *A-1 Operator Club*. Membership is available only by nomination by two present club members, and is awarded in recognition of overall top-notch operating ability and courtesy shown by a deserving operator. The only way to attain this honor is by consistently practicing good operating methods and by demonstrating a courteous attitude toward fellow operators. There is no way to "apply" for an award—in fact, it is said that anyone who openly seeks this award is almost certain to *not* receive it!

The *BPL* (*Brass Pounders League*) award is given to operators who handle a sizeable number of formal messages over a period of time. A card-sized certificate is given for handling a required number of messages in a month, and a BPL medallion is awarded after three certificates have been won.

The BPL award originally began in the days when virtually all traffic was handled by cw, hence its name. However, it is also available to operators using phone or rtty.

Another recognition of public service communications is the *Public Service Honor Roll* listing. Credit points in this monthly listing are given for such activities as participation in a public service net, operating as a net control station, serving as a liaison station between two nets, and serving as a net manager.

ARRL FIELD ORGANIZATION

For any specialty which a ham chooses, there are hundreds of other hams interested in the same things, so nearly every ham gravitates toward one or more groups of fellow hams with similar interests. Many organized groups of amateur radio operators have been formed around the various specialized amateur operating interests.

Tying many of these special operating interests together is the field organization of volunteer hams, sponsored by the ARRL, to self-administer their own operating activities.

The "chief honcho" in each area of the country in this organization is the *Section Communications Manager* who is elected to a two-year term by ARRL member hams in his section. In many cases, a state comprises a section, but in more densely populated areas, a state may be divided into two or more sections.

Within each section are numerous field organization appointments, each reflecting a prime interest of the individual operator. Each appointment is made by the Section Communications Manager (SCM) upon application by the individual amateur operator.

These appointments, and their functions are:

Section Emergency Coordinator—Supervisor of emergency communications preparedness organization within the section. This organization is called the *Amateur Radio Emergency Corps* (*AREC*).

Emergency Coordinator—Serves under the SEC to supervise emergency communications activities of the AREC in individual communities within the section.

Route Manager—Manager of a cw traffic net (see Chapter 7).

Phone Activities Manager—Manager of a phone traffic net (see Chapter 7).

Official Relay Station—Individual amateur station which participates regularly in cw or rtty traffic handling.

Official Phone Station—Individual amateur station which participates regularly in phone traffic handling.

Official VHF Station (formerly called Official Experimental Station)—Vhf amateur who works regularly on experimental electronics projects related to vhf operating and reports results obtained, or who participates in vhf traffic work.

Official Bulletin Station—Amateur station which regularly transmits news bulletins received from ARRL Headquarters.

MONITORING ACTIVITIES

Two highly specialized activities which involve *monitoring* instead of *operating*, and in which a small but active group of amateur operators are involved are the *Official Observer* corps and the *Intruder Watch*. While amateurs in these two ARRL programs generally are active on the air in other activities, these two particular activities involve highly specialized listening instead of transmitting.

The purpose of the Official Observers is to notify other amateurs of any FCC rule violations which are detected so that the amateurs involved may correct the problems without being officially cited by the FCC. Most of these violations are of a technical nature, such as out-of-band signals (especially out-of-band harmonics), chirp, key clicks, inadequate power-supply filtering, excessively broad phone signals, and the like. Some other violations which are sometimes detected include improper operating procedures, such as improper call sign identification and use of improper language.

Whenever an Official Observer (called an "OO") detects a violation, he sends a post card to the station operator involved. No mention of the violation is made on the air, to avoid embarrassment to the operator involved, and no report is made to the FCC. The OO program is strictly a "self-policing" operation by amateurs and for amateurs, and has no connection with FCC monitoring and enforcement activities. However, the FCC has repeatedly praised the OO program for its service in helping amateur operators clean up their own problems without the necessity of government enforcement.

The Intruder Watch is a program started a few years ago to reduce the illegal interference with amateur operating by commercial stations which sometimes move into the ham bands. Some of these are foreign propaganda stations, and some are commercial or government communications setups in various areas of the world.

Members of the Intruder Watch, which is sponsored by the ARRL, monitor amateur bands and identify nonamateur stations which they hear interfering with amateurs. These reports from individual Intruder Watchers are sent to ARRL Headquarters, where they are summarized and sent to the FCC. On the basis of these reports, the FCC and the U.S. State Department has exerted pressure on foreign governments and commercial interests and caused many of these interfering stations to be moved to other frequencies or put off the air.

It should be pointed out that some amateur frequencies are shared with nonamateur stations in some parts of the world, so some of the nonamateur interference you hear on the ham bands is there legally—but not all of it, by any means.

Chapter 5

Observing Frequency Boundaries

Regulations governing ham radio operations specify only that hams must remain within the amateur bands, and that they must remain inside specific band segments appropriate to their mode of operation and class of license. That is about as simple as any frequency requirements can be, and allows amateurs much more freedom than is permitted in commercial operations, where stations are restricted to specific frequencies which they must attain within very narrow tolerances.

However, it is still important for all amateurs to not allow this freedom to lull us into an easy-going attitude and stray from our individual part of the spectrum. This can happen easier than we sometimes imagine. Since the reintroduction of the principle of incentive licensing, whereby sections of some ham bands are restricted to use only by certain classes of licensees, it is even more important that we keep in mind the boundaries within which each of us may operate, and understand how to be sure to not overstep these boundaries.

Violation is not something to create a screaming panic in the radio shack, but it is something which should be taken seriously and be avoided with great care. If an FCC monitoring station hears your signal out of the band, or out of a band segment appropriate to your mode of operation or your class of license, you will be sent a *Notice of Violation* giving details of the incident. You are required to reply, stating what steps you are taking to prevent a recurrence of the violation. Occasionally a smarty decides he will foil the federal machinery by ignoring the FCC notice. This is always a mistake because when Uncle Sam loses patience with such a person, he merely suspends or revokes that fellow's license to operate.

Amateurs have available many modes of operation today. This station is equipped for cw, ssb, f-m, RTTY, and ATV. In some cases, only part of a given amateur band may be used for certain transmission modes, and amateurs must be able to measure their operating frequencies well enough to be sure they stay inside these subbands, as well as inside the total amateur band.

By answering the violation notice promptly and giving a sensible, practical explanation for how the violation occurred and stating what will be done to prevent it occurring again, you will end the entire affair—unless the violation is repeated. If that happens, you could be called forth to answer personally for your actions.

Amateur radio operators may take pride in the fact that a relatively small number of violations occur in the ham bands, and that most of these are for minor "accidents" that probably happen only once in a ham's lifetime. Much of the reason for this good record is that amateur radio in the United States has always been a self-policing operation. Hams, out of a friendly, cooperative spirit toward one another, generally notify one another of violations they detect, so that the individual operator involved may correct his own mistakes before he gets into trouble with the FCC. Helping do this self-policing of the ham bands are the "Official Observers," discussed in the previous chapter.

OUT-OF-BAND HARMONICS

An easy way for an amateur to put a signal outside the amateur bands is by unknowingly radiating a *harmonic*. A harmonic is any multiple of a fundamental frequency. All oscillators produce harmonics, but with proper design and operation of the transmitter, these extraneous signals are not radiated by the antenna.

Sometimes, though, a harmonic is radiated, and this is a potentially dangerous condition. If the harmonic falls within a ham band, it will cause interference to other hams, which is bad enough, but if the harmonic falls outside a ham band, on some commercial frequency, it can cause unwarranted interference to commercial communications, some of which have life or death importance. This is one of the types of rule violations which Official Observers are constantly looking for, so that they may notify the errant operators and enable them to correct the problem before they get in trouble with the FCC.

Searching for harmonic radiation is difficult for a ham to do alone, since he usually will hear harmonics of his transmitter oscillator in a receiver in the shack even though the harmonics are not being radiated by his antenna. A *wavemeter*, which used to be a standard ham shack accessory can be helpful in tracking down transmitted harmonics since it is less sensitive to such signals than a receiver, so is less likely to give false indication of harmonic radiation. However, modern commercially built transmitters, which usually are free of harmonics (but not always) have caused the wavemeter to be neglected nowadays.

Enlisting the aid of a ham friend who lives a few miles away and has good receiving gear is about the best way to check for harmonics. Tell him what frequency you are operating on, so he can tune a receiver to harmonics of that frequency while you transmit.

FREQUENCY MEASUREMENT

Even if we have a clean signal, with no harmonics, we still must keep this signal within the band on which we are operating. The FCC does not require that we be able to measure our operating frequency precisely, as it does with commercial operations, but we must have the ability of knowing that this signal is within the appropriate amateur band.

The first consideration in doing this is that we always keep in mind the fact that we cannot measure our frequency *exactly*, so we must make allowance for this fact. This is a basic lesson faced by every beginning ham with a Novice license, but it is well for all of us to keep this lesson in mind all the time.

Take the example of the Novice on 40 meters going to the local radio store to buy some new crystals. If he has been studying the signals he hears on his receiver, he probably has discovered that often there is less congestion near the edge of the Novice band segment than toward the middle. This means he may be heard better if he operates near the edge because there is less likelihood his signal will be covered by other operators on nearby frequencies.

The 40-meter Novice band is from 7150 to 7200 kHz. So, should a Novice wishing to operate near the bottom end of this band segment buy a crystal marked 7151 kHz? Not unless he has an exceptionally precise way of measuring the frequency of this crystal *when used in his transmitter!* The fact of the matter is that a crystal seldom operates *exactly* on the frequency which is marked on the crystal holder. There are two factors responsible for this, and they must be considered when buying crystals near a band edge.

One is the tolerance of the crystal calibration—a crystal may be marked 7151 kHz, but this is no assurance that it is cut to *exactly* this frequency. The amount by which it may vary is the crystal "tolerance," and usually is marked on the crystal holder, or shown on the store display or in the catalog listing. A tolerance of .05% is typical.

Let's suppose the crystal marked 7151 kHz actually oscillates .05% lower than the indicated frequency. This

FCC rules require that an amateur use a means other than transmitter dial calibration to check his operating frequency. Use of a reliable receiver with a crystal calibrator which has been checked against a known-frequency signal is most common of several ways to be sure the transmitted signal is in legal territory.

would be 7151 − 3.58 kHz, which figures out to be 7147.42 kHz, which is out of the Novice band. So anyone buying a crystal, whether Novice or not, needs to keep in mind the crystal tolerance and calculate what the actual oscillator frequency *could be* at the highest and lowest frequencies within this tolerance range.

A second factor to remember is that the frequency at which a crystal will oscillate may vary several hundred hertz up or down, depending on the oscillator circuit in which it is used. You might determine the frequency at which a particular crystal oscillates in your transmitter, then try it in a friend's transmitter and discover it oscillates at a slightly different frequency, due to differences in the two oscillator circuits. So, the wise operator selects crystals that will keep him safely within the band edges even if the "worst conditions" of deviation from the frequency marked on the holder do occur.

If a crystal is needed to be as near to a specific frequency as possible, such as for a net operation, then it is best to order a crystal cut to the desired frequency from a manufacturer that will make it for the specific oscillator circuit in the transmitter in which it will be used. These cost a little more, but the price is well worth it for the peace of mind and operating convenience that results, if this precision is needed for a particular operation.

MODULATION AND KEYING BY-PRODUCTS

Another important fact to keep in mind while operating is that keeping the carrier frequency inside the band does not necessarily give assurance that all components of the radiated signal also will be within the band. Among items to take into consideration in this regard is the bandwidth of the radiated signal. If any portion of the signal is radiated outside the band, it is considered a rule violation, even though the carrier which produced the signal is inside the band.

Take for example a phone signal, either ssb or a-m. When the carrier frequency is modulated, additional frequencies above and below the carrier frequency are produced having a direct relationship to the audio frequencies in the modulating signal.

Let's say the carrier frequency is 3.98 MHz, just inside the top of the 75-meter band, and that it is modulated with an audio signal containing frequencies up to 3 kHz, which would be typical of a voice spoken into a microphone.

This modulation will cause rf frequencies up to 3 kHz above and below the carrier frequency to be produced. These frequencies, called *sideband frequencies,* would, in this case, extend from 3.95 MHz, to 4.01 MHz. You see the upper sideband frequency would extend out of the 75-meter ham band. Consequently, transmission of this signal would be a violation of the FCC rule on allowable frequencies, even though the carrier which produced the signal was inside the band.

In a-m operation, the carrier and both sidebands are radiated, while in ssb operation the carrier and one sideband are suppressed, with only a single sideband radiated under optimum conditions. So, it is possible to operate ssb closer to the band edge than a-m, *if* you are very sure the carrier and unwanted sideband are adequately suppressed.

Of course it is also important to remember which sideband is nearest the band edge, and stop to think for a second whether this is the sideband you are operating on, or whether you are transmitting the opposite sideband. For instance, if you are operating on *lower sideband,* you could operate closer to the *top* end of a phone band, and if you are operating on *upper sideband,* you could operate closer to the *bottom* end of a phone band, without radiating a portion of your signal out of the phone band; that is, *if* you are fully confident you have *full* opposite-sideband suppression. The answer to that "if" could come in an envelope from an FCC monitoring station.

One little slip in tuning of your rig and you would have a portion of your signal outside the band edge. Consequently, it is wise for the ssb operator to assume his unwanted sideband is not totally suppressed when he selects an operating frequency near a band edge.

On cw, it is also possible to have a component of your signal outside the band, if you are unfortunate enough to have key clicks. Although they are what you might think of as "tiny bits of rf" that could not really matter much, they actually can create damaging interference on non-amateur frequencies.

If you encounter someone on the air with key clicks and tune to each side of his signal, you will still hear these annoying clicks, even though you cannot hear his carrier. Naturally, if he was operating near enough to a band edge for some of these clicks to be radiated outside the band, he would be guilty of out-of-band operation, even though his carrier itself was in perfectly legal territory.

These examples point up important reasons for all amateurs to strive to have clean signals. It is not just so they will sound good on the ham bands, but also so they will not cause interference outside the ham bands. Cw key clicks, insufficient suppression of carrier or unwanted sidebands in ssb rigs, and parasitic oscillations in any rig can produce signals that are radiated outside the ham bands, no matter if the carrier itself is properly within the band.

Incidentally, it is good to remember that the FCC has ruled that an amateur must not only have all components of his signal inside a ham band, but he must have all components inside the particular portion of the band for which he is licensed. For example, a General class operator must not allow a modulated subband of his signal, or a burst of key clicks, to be radiated on frequencies restricted to Advanced or Extra class licensees. Even though the carrier is in the proper subband for the class of license held, and the entire signal is inside an amateur band, it is still a violation to allow by-products of the signal to appear outside the appropriate subband.

This rule also means it is illegal for a portion of the modulated sideband of a phone signal to fall outside a phone band, even though the carrier frequency is in the phone band and the entire signal is inside an amateur band. So, a U. S. amateur operating lower sideband near the bottom of a U. S. phone band could be cited by the FCC for allowing part of his sideband to be radiated outside the lower limit of the phone band, even though it still would be within the overall amateur band.

DON'T BELIEVE THE DIAL!

In radio operating, just as in many other activities, it is well to maintain a questioning attitude. This is especially appropriate in determining frequencies on which equipment is operating. The commercially built equipment which is so commonly used in ham radio today is so nice looking, it carries with it an aura of respectability and precision. And for the most part it is pretty well made, but don't make unwarranted assumptions about the precision of something so important as the frequency dial on a transmitter or receiver. Certainly they will be very *close* to being right, but the question that should always be kept in mind is: "How close?"

Can you be positive the calibration on your gear is correct to within 1 kHz? Or 500 Hz? How do you know? Would you bet your license on it?

A little checking against a good frequency meter usually will reveal that the VFO dial is very close to being correct, but is seldom absolutely correct along its full scale length. It is virtually impossible to mass-produce frequency scales that are exact along their full length without going to expensive manufacturing techniques. It really is not necessary anyway for most amateur operations, because there are less expensive ways to locate a frequency with sufficient precision for nearly all amateur needs. Also, most frequency dials on amateur gear are not designed to be read with really high precision. The width of the pointer or of scale calibration marks could easily throw you off several hundred hertz.

You really do need methods other than direct dial reading, anyway, for transmitter frequency adjustments because there is a little item in the FCC rules that states: *The licensee of an amateur station shall provide for measurement of the emitted carrier frequency or frequencies and shall establish procedure for making such measurement regularly. The measurement of the emitted carrier frequency or frequencies shall be made by means independent of the means used to control the radio frequency or frequencies generated by the transmitting apparatus and shall be of sufficient accuracy to assure operation within the amateur frequency band used.*

This would be a wise precaution even if it wasn't in the rules because think what could happen if your transmitter VFO dial slipped a couple of gear teeth, or a trimmer capacitor in the oscillator got bumped out of position when you changed a tube. You could go on the air on a frequency far from where you thought you were, because you had no independent means of checking your frequency.

This independent means does not need to be an elaborate laboratory frequency meter; a reliable means that will assure that you remain inside the amateur bands, and the appropriate band segment, is all that is required by the FCC. In achieving this, you generally will wind up with something that will measure your frequency anywhere across a ham band with a practical level of precision.

A typical ham on the high-frequency bands uses his receiver, with a crystal calibrator accessory, to perform the measurement function. It is simple, direct, and when properly used is well within the frequency tolerance required.

USING A CRYSTAL CALIBRATOR

Most modern ham receivers have a 100-kHz crystal calibrator built in, or available as a plug-in accessory. This oscillator will provide harmonics each 100 kHz all the way across the hf bands, so as you tune your receiver across a band, you can hear a signal from the oscillator at every 100-kHz point across the dial. To be useful, this oscillator must be adjusted so it is on 100 kHz as precisely as possible. Generally there is a small trimmer capacitor in the oscillator which may be adjusted with a screwdriver to change the oscillator frequency slightly. This adjustment, if needed, is made so the oscillator produces a true zero beat with a signal from WWV. This should be repeated every few months, because the 100-kHz oscillator may drift off frequency slightly due to changes in temperature and humidity, and to component aging.

The crystal calibrator in a receiver generally is connected so the output of the oscillator is fed to the antenna input of the receiver, so whenever the crystal calibrator is turned on, it will feed its 100-kHz harmonics to the receiver just as if they were incoming signals at the antenna. Usually a receiver which has a calibrator also has an adjustment knob on the front panel so the dial scale or pointer may be moved without changing the receiver tuning.

With such receivers, the procedure used to calibrate the frequency dial is this: switch on the crystal calibrator and the receiver's BFO, and rotate the tuning knob until you find the 100-kHz harmonic *nearest* the frequency you intend to operate, or listen, on. Slowly turn the tuning knob until you have this signal tuned in for maximum strength. Then adjust the BFO pitch control until you obtain a zero beat. Now, without disturbing the receiver tuning, carefully turn the dial pointer adjustment knob to get the pointer exactly on the 100-kHz harmonic point on the scale. Now the receiver dial is corrected for this point on the dial, and should be quite close to being correct for a moderate distance on either side of this point.

If you tune the receiver to another 100-kHz point on this same scale, and again turn on the crystal calibrator, you may find the dial indication is no longer exactly correct. This is because of the difficulty of building a receiver that is exactly linear across its entire tuning range.

The fact that a slight readjustment, using the 100-kHz crystal calibrator, may be necessary when you move a considerable amount across a band is one of the things to keep in mind when determining what frequency you or someone else are on. If you ignore or overlook this need for calibrating the dial in the portion of the band in which you intend to operate, you can easily be wrong by a kHz or more on many ham band receivers, and by more than that on a general coverage receiver.

A crystal calibrator which also will produce 25-kHz harmonics is useful, since it provides the necessary additional frequency check points within the ham bands which set apart certain subbands for Advanced and Extra class licensees.

Such units are available factory-made now, and several designs for home-built units have appeared in electronics magazines.

For use throughout the ham bands, a commercially built frequency meter provides the most flexibility, since it can locate *any frequency* on a band, rather than just 100-kHz and 25-kHz mulitples. Many hams have acquired such meters from military surplus or commercial sources.

Regardless of the means used, it is important to understand the proper way to calibrate a dial using a calibration signal, such as from a crystal calibrator or a precision frequency meter. This is often misunderstood, and can lead to heated differences of opinion (to put it mildly) among operators as to just who is on what frequency.

ADJUSTING BFO PITCH

To adjust the dial calibration properly, it is necessary to understand the meaning of a true zero beat, and of the function of a BFO. If you switch on the BFO by switching the receiver mode switch to "cw," and then switch on an external signal, such as a crystal calibrator, transmitter VFO, signal generator, or frequency meter, and then tune the receiver to near the frequency of this signal source, you will hear a tone in the receiver.

This tone is the "beat note," or audio heterodyne frequency produced by the external signal mixing with the BFO in your receiver, which explains why it is called a *beat-frequency oscillator*. Now, if you want to tune the receiver to exactly the same frequency as the external signal source, you must adjust the tuning control or the BFO pitch control until this tone drops in pitch so low you can't hear it —this is the *zero-beat* point.

Actually, since your ears cannot hear frequencies below a hundred Hz or so, you can't be sure of getting a perfect zero beat by ear alone, but you can get close enough to it to suffice for all practical purposes. This zero beat is the reference point to use for greatest precision when adjusting your transmitter to the same frequency as an incoming signal, and for setting a receiver or a signal generator or a crystal calibrator to a particular frequency with the most precision you can muster.

When you wish to receive a cw or ssb signal, however, you must tune the BFO (or product detector in the case of a modern ssb receiver) off to the side of zero beat slightly in order to make the transmission audible and understandable. It is when doing this that some operators fail to get their receiver tuned to the frequency they think they have. For example, if you tune a cw signal to zero beat, then retune so you can hear the note, you actually are tuned slightly to one side of the carrier frequency. So, if you read the dial after doing this, it will indicate a frequency which is off the frequency of the signal you are hearing by an amount equal to the frequency of the tone you are hearing in the speaker or headphones.

To overcome this problem, many receivers have their BFO pitch controls, or product detector controls, calibrated with marks indicating the position to be used for zero beat and the positions to be used for cw and for upper- and lower-sideband reception.

The exact method used with these controls will vary from one manufacturer to another, but armed with the forgoing precautionary information, you will be able to understand the operator's manual and put it to proper use in receiving the various types of signals and knowing how to read the dial in each case.

The principle involved is this: To tune in the signal and be able to read the proper frequency from the dial, it is necessary to set the BFO pitch control to its center or zero-beat position, turn the tuning knob until the signal you are hearing produces a zero beat with the BFO, then reset the BFO pitch control for the most pleasing audio tone *without touching the tuning control.* If done in this way, the frequency dial will show you the frequency of the signal you are hearing; assuming, of course, that you have first calibrated it properly.

An alternate method, which is faster, but not quite so precise, is to leave the BFO pitch control set where it produces the most pleasing note from an incoming signal, and tune in the crystal calibrator so it produces this same note. Then set the dial adjustment so the scale pointer is on a 100-kHz point. Now when you tune in a cw signal with the same pitch as you used in calibration, the dial will give you the proper indication. This method obviously can be done quicker, but its precision depends on your ability to match audio tones strictly by ear, so there likely will be a little error in your dial indications.

Methods similar to those described here for cw may be used for tuning a-m and ssb signals, with the principal difference being that on a-m you will tune for a zero beat with the a-m carrier, while on ssb you will tune for the most pleasing audio characteristic of a received ssb signal. In either case, when you can hear the audio modulation clearly, you have the receiver BFO beat against the carrier of the station to which you are tuned. (In the case of ssb, the actual carrier is not transmitted—this is in effect supplied by the BFO of your receiver, but the end effect is the same as if there were a transmitted carrier present.) On cw, however, the BFO must be shifted slightly from the carrier so you will be able to hear the tone when copying the code, but the BFO should be *zero beat* with the carrier by means of the BFO pitch control (*not* the receiver tuning control) when reading the signal frequency.

Once you have mastered the procedures appropriate to your particular receiver for reading the frequency, it is a simple matter to apply this procedure to determine what frequency your transmitter is on, since you can hear the signal from the VFO in your receiver the same as you hear another station on the band.

If the signal from your VFO is not too strong in your receiver, you can use it instead of your BFO to set your transmitter to the frequency of a station you are receiving, just by tuning the VFO until the received station's audio sounds normal. If the signal from your VFO is strong enough to overpower the received station in your receiver, this won't work, so you will need to follow the previously described method of setting the BFO to zero beat with an a-m carrier, or so it recovers the audio in an ssb signal, then set the VFO to zero beat with the BFO.

These procedures may sound complicated at first, but they will become second nature to you after you have followed them several times and acquired a feel for what is happening. The methods described enable you to determine with fair accuracy the frequency of a station you hear in the band, and of course at the same time they enable you to determine the frequency on which you will transmit; and they utilize methods entirely independent of the transmitter VFO calibration, which is the important FCC requirement previously stated. In fact, they would work even if there were no dial on your VFO at all.

Chapter 6

Utilizing Propagation Effects

Most contacts made by amateur radio operators on the high-frequency (hf) bands (80 through 10 meters) and a small portion of the work done on vhf bands (6 and 2 meters) are made possible by refraction of radio waves by the ionosphere. The ionosphere is a layer of charged particles which surrounds the earth, much like the skin of an orange surrounds the meat of the orange, except that in the case of the earth there is a gap between the earth's surface and the underside of the ionosphere.

The thickness of the ionosphere varies with varying action on it by the sun, and this variation in thickness changes the effect of the ionosphere in refracting or bending radio waves back to earth. In very simplified form, this is what causes certain amateur bands to have different characteristics at night than they do in the daytime, and different characteristics in the summer than in the winter. The ionosphere's ability to bend radio waves back to earth also varies with the 11-year sunspot cycle, and this is the reason that your favorite amateur band may be better (or worse, depending on what you want to accomplish) this year than a year or two ago.

A detailed study of the action of solar activity on the ionosphere and the resulting effect on radio propagation can easily fill a book in itself, but it is possible to draw up certain rules of thumb which will easily serve to predict approximately what distances you can cover on any particular band at a given time of day and time of year.

EFFECT OF IONOSPHERE THICKNESS

The main key to remember is that increased exposure to the sun, or increased solar activity in the form of sunspots, causes the ionosphere to become thicker. In conjunction with this, it should be remembered that radio waves entering the ionosphere at an angle are bent, much the same as light rays entering glass or water at an angle are bent. As the ionosphere grows thicker under the influence of the sun, radio waves entering it have a greater distance within which they may be bent. If this distance is great enough, the radio waves are bent far enough that they are returned to earth.

By having solid understanding of propagation effects and knowing how to use them to accomplish specific operating goals, amateurs can accomplish much without elaborate equipment. This station is largely homemade, yet it is capable of matching the performance of more elaborate rigs through scientific approach to operating.

It has been found that as the frequency of a radiated signal is increased, a greater thickness of the ionosphere is required to bend the signal sufficiently to cause it to return to earth. In terms of typical amateur bands, the ionosphere must be thicker to return a 15-meter signal to earth than a 20-meter signal, for example. Anyone who has studied ionospheric propagation in all its excruciating detail will wince with pain at this extremely simplified description. However, for the radio operator who is primarily interested in determining what band to use to communicate over a given distance at a given time, this simple generalization will help figure out the answer with a minimum of effort.

The factors to be considered in practical, everyday radio operating is the time of day, time of year, and phase of the sunspot cycle. By considering these in the light of how the sun affects the ionosphere, you can come up with a pretty good "educated guess" of what you will hear when you turn on your receiver, and of course what distance you can work with your transmitter. To do this, it is only necessary to think for a moment how the time of year and time of day affects the amount of time during a 24-hour period that the sun may influence the ionosphere between your location and the location you want to work by radio.

A few examples should make this clear. In the daytime, the sun is above the horizon and so its rays are striking the ionosphere over your head and in all directions from your location. So, you would rightly expect the ionosphere over your general part of the world to become thicker in the daytime. At night, with the sun on the far side of the earth, the ionosphere over your general part of the world is shaded from the sun, so the ionosphere will slowly disperse, becoming thinner.

In the summertime, what change takes place? The sun is above the horizon in your part of the world for a longer period each day (in the northern hemisphere, that is—just the opposite is true in the southern hemisphere). So, the ionosphere in all directions from your location will receive rays from the sun for a longer period of time than it will in the winter, when the days are shorter. Consequently, thickness of the ionosphere increases in the summer in the northern hemisphere because there are more hours of sunlight during which the ionosphere builds up, and fewer hours of darkness during which it can disperse. (This happens during the winter in the southern hemisphere.)

Now, thinking back to your experiences on various ham bands, you will recall you can work long distances on 20 meters, for instance, better in the daytime than at night, and better in the summer than in the winter. The reason now becomes obvious—the ionosphere is thicker at those times, and consequently more able to bend your 20-meter signal back to earth at some distant point.

Amateur contest operators, such as this operating team at a radio club Field Day setup, can greatly improve their scores by understanding how propagation effects will influence distances they can cover on each band. By concentrating on each band at the time of day or night when it will be most useful, they make many more contacts than can be made merely by sticking doggedly to one band.

A third factor that is prominently involved is the sunspot cycle. Over approximately 11-year periods, the number of sunspots build up, then decrease, following a fairly predictable pattern. At about 11-year intervals, we experience a "sunspot maximum," when numbers of these magnetic tornados on the surface of the sun reach their peak intensity and peak numbers. The number and intensity of sunspots intensifies the effect of the sun's rays on the ionosphere. The result is that the thickness of the ionosphere and consequent ability to bend higher-frequency radio waves back to earth increases with increasing sunspot activity.

Remembering that as the ionosphere grows thicker, the frequencies which it will bend back to earth become greater, you can see why if a given radio contact could be made on 20 meters at a certain time of day and time of year when sunspots are at a minimum, this same radio contact might be made on 10 meters, or possibly even on 6 meters, during intense sunspot activity, due to the increased ability of the ionosphere to refract higher frequency signals back to earth.

IONOSPHERIC ABSORPTION

To make the picture complete, we also need to consider *absorption* of certain radio waves by the ionosphere.

This occurs in the lower layers of the ionosphere and causes certain frequencies at certain times to be *poorer* for longer distances, while at the very same time, other frequencies are *better* for the same distances.

Again in very simplified form, it works like this: when the ionosphere is quite thick, lower-frequency signals are absorbed by the bottom layer of the ionosphere, while at the same time higher-frequency signals penetrate this lower layer and continue upward into the ionosphere far enough to be bent back to earth.

For example, at a time when 15- or 20-meter signals are being bent back to earth and producing strong signals several thousand miles from the transmitter, signals on 40 and 80 meters will be absorbed by the lower layer of the ionosphere. This condition would exist during daytime, especially in the summer, and to an ever greater extent during periods of high sunspot numbers.

However, as the ionosphere thins out, there is less absorption of lower-frequency signals, such as on 40 and 80 meters. These signals then can penetrate farther into the ionosphere. Even though the ionosphere may not be thick enough at this time to bend 20-meter signals back strongly, it will bend 40- and 80-meter signals back to earth. So, during times when the ionosphere is thinner (such as at night, in the winter, and during low sunspot activity), it is possible to work greater distances on 40 and 80 meters than on 15 or 20 meters.

EFFECT ON AMATEUR BAND PROPAGATION

Putting all this information together and applying it to the hf amateur bands, we can form several general conclusions that will be helpful in amateur activities.

For one, we can expect to work longer distances on 10, 15, and 20 in the daytime than at night, and we can expect to work longer distances on 40 and 80 at night than in the daytime.

We can also expect to work longer distances, *for more hours at a time* on 10, 15, and 20 in the summer than in the winter, and we can expect to work longer distances for more hours at a time on 40 and 80 in the winter than in the summer.

And finally, we can expect to work long distances on 10, 15, and 20 more readily during years of high sunspot numbers, and we can expect to work long distances on 40 and 80 meters more readily during years of low sunspot numbers.

With these rules of thumb, we can plan the best band to use to have the best signal and the least interference for whatever it is we want to accomplish. A few examples will illustrate how to do this:

(1) Suppose you live on the west coast and you want to make a schedule with a friend in the eastern part of the country. If you can do it during the daytime, your best bet would be 15 or 20 meters. You could do it easily on 10 meters, and possibly on 6 meters if it happens to be a year in which there is a large amount of sunspot activity. However, if you need to make the schedule at night, you will probably do better on 40 or 80 meters, although you might be able to do it at night on 20 meters during a period of high sunspot activity.

(2) Suppose you want to organize a traffic net, which will meet in early evening, and will cover an area of about 300 miles radius. Your best bet will be 80 meters (or 75 meters if it is to be a phone net). However, during periods of low sunspot activity, you will get a lot of interfering signals from more distant areas after sunset. So, as the sunspot numbers decrease, you might want to move the net time up into late afternoon to minimize this interference caused by distant signals in "skipping" due to not being absorbed in the lower level of the ionosphere during low sunspot activity.

(3) Suppose you want to work DX, and build up a collection of contacts aimed at acquiring a DXCC certificate. If you can make your radio contacts in the daytime, 15 or 20 meters should be a good bet, and 10 meters will be good if the sunspot numbers are fairly high. You will also find 20 meters and possibly 15 meters good at night if there are a large number of sunspots. However, if the sunspot

number is fairly low, your best bet will be 40 meters at night, and possibly 80 meters quite late at night.

(4) Suppose you want to organize a local area "rag-chew" get together to meet late on Saturday nights. During periods of medium to high sunspot activity, your best bet will be 80 meters, although you will have a moderate amount of interference from stations a few hundred miles away, especially in the summertime. During periods of low sun-spot activity, especially in the winter time, your best bet would be 10 meters, since it will be virtually dead to any signals more than a few miles away.

There are many other combinations of operating activities and operating conditions that will occur to you, now that you can see what causes the various propagation effects you will experience. The things to keep in mind to achieve the most success with whatever type of operating you are interested in, is the distance which you can expect your signals to travel and the possibilities for hearing other signals which may create interference for you.

In some cases, the type of antenna will help or hinder what you are trying to accomplish. For long distance work, a directional antenna helps, because it concentrates your transmitted signal and improves your reception in one direction, while minimizing reception of signals from other directions. For short distances, this can be a handicap if you want to work more than one station simultaneously, because having the antenna pointed at one may not produce the best signal from another.

For longer distances, an antenna with a lower angle of radiation will help, whether it is directional or not, because it will cause the signal refracted in the ionosphere to return to earth farther away. For shorter distances, an antenna with a higher angle of radiation will improve your signal and your reception of stations at lesser distances, while minimizing reception of interfering signals from more distant areas.

Chapter 7

Message Handling

Handling messages on behalf of non hams is one of the backbone activities of amateur radio. It dates back to the very early days of amateur radio operating, and has won worldwide recognition for hams because of the frequent humanitarian service provided. In times of emergency such as following earthquakes, floods, explosions, train wrecks, forest fires, and other disaster situations, thousands of amateurs have helped relieve suffering and mental distress by handling messages for rescue workers and for victims and their families.

Less spectacular but just as important is the routine message-handling service provided daily by amateurs for the enjoyment of operating and as practice to remain prepared for unexpected emergencies. This message-handling service is recognized in the opening section of the FCC amateur rules which lists functions of the amateur radio service and includes "service to the public as a voluntary noncommercial communication service. . . ."

Because of the wide variety of types of equipment and operating skills available among a group of radio amateurs, and the capability of operating on frequencies suitable for blanketing a small area or reaching around the world, hams are in an ideal position to provide message service for all sorts of noncommercial purposes. This communication usually is done through organized groups of hams who form a traffic "net" (meaning "network"), but it may be provided directly between two stations, on either a prearranged schedule or an informal contact if desired.

Nets have the advantage of being set up to handle traffic from anywhere to anywhere through a series of scheduled relays, whereas direct point-to-point traffic handling usually must be scheduled in advance to be certain contact can be made with a station in the location desired. In either case, the objective is to provide speedy, accurate transmission and delivery of brief messages anywhere in the United States, and in some cases between the United States and foreign countries.

The word *PICON* was coined years ago to symbolize the term "public interest, convenience, and necessity" as a basic foundation of amateur radio, and an important justification for our frequency assignments. Two important ways in which amateur radio serves the public interest, convenience, and necessity is as a message-handling service and as a technical training ground for electronics enthusiasts.

It is not intended that every amateur radio operator should busy himself nightly with handling of messages if he is not interested in it, since there are other equally important facets to ham radio, and all hams are not expected to be interested in the same things. However, it has often been pointed out that every ham has an obligation to his community and to his fellow hams to at least *understand how to participate in message-handling activities* so he will be able to perform this function if circumstances require it.

Some amateur operators make the handling of messages their principal operating activity, and find it a very rewarding experience, even during nonemergency times when all traffic is routine. Through handling dozens of these "happy birthday" and "having a wonderful vacation" messages, they sharpen their operating skills and maintain regular contact with other traffic-handling operators, so they are always prepared to deal with unexpected urgent situations which otherwise could create large scale confusion on the bands. When an emergency occurs, it is too late to get organized—the preparation must be done in advance and practiced frequently, so that the message-handling organization is all ready to swing into action when an emergency does occur.

Traffic handling as a public service is a satisfying activity pursued by many amateur operators. By maintaining operation of daily traffic nets, these operators are continually organized and practiced in preparation for handling messages in event of any emergency for which their skills are needed.

Since all amateur radio operating must be strictly noncommercial, naturally it is forbidden to accept any payment for the handling of messages. Most dyed-in-the-wool traffic-handling amateurs are insulted by any offer of payment anyway—they find the pleasure and satisfaction of knowing they are putting their radio operating skills and equipment to work as a public service to be ample reward for the time and money it costs them to provide this service.

Messages may be from one ham to another, or from a nonham to another ham or nonham—it makes no difference. Since many messages sent by one amateur to a second amateur are sent on behalf of a person who is not a ham, such traffic generally is referred to as "third-party" traffic.

Messages may be handled between any two or more operators in the United States, and between U. S. operators and operators in certain foreign countries. Of course U. S. hams may talk informally on the air with hams of nearly all other countries, and may exchange information of a personal nature or related to their radio interests at any time. However, not all countries permit the exchange of messages on behalf of third parties (nonhams) with other nations. A list of countries permitting such messages appears later in this chapter and in the appendix.

Certain special rules are adopted from time to time covering handling of messages on behalf of servicemen stationed overseas and in connection with specific international events involving amateurs—more on that later in this chapter. Within the United States and Canada, where the vast majority of all messages are handled, there are no restrictions other than the standard requirement that the message content be noncommercial in nature.

NET ORGANIZATION

To facilitate the handling of messages, a large number of traffic nets have been established throughout the United States and Canada. These nets provide a systematic organization for sending messages anywhere in the two countries. They operate on regular schedules and on preannounced frequencies, so anyone wishing to send a message can easily find out the time and frequency on which to do it.

Traffic nets, sometimes called "public service nets," are set up strictly for handling of formal messages. You also will find a number of "social type" nets or rag chew nets on the bands. These operate mainly for the enjoyment of informal chatter between hams with similar interests. Some of these also handle a little traffic, but that is not the primary interest of their members. In some areas, there also are technical nets, composed of operators with common technical interests who have selected a time and frequency to meet and discuss projects of mutual interest.

It is well to keep these other types of nets in mind when you get interested in traffic handling, so you are forearmed with knowledge of which type of net to look for.

Many of the public service traffic nets operate as an official part of a national system set up to handle messages to and from any part of the United States and Canada. This organization is known as the National Traffic System (NTS), which is sponsored by the ARRL. It is the backbone of a nationwide complex of traffic operators.

It would be impractical to try to put traffic operators over the entire country into a single colossal net, so the

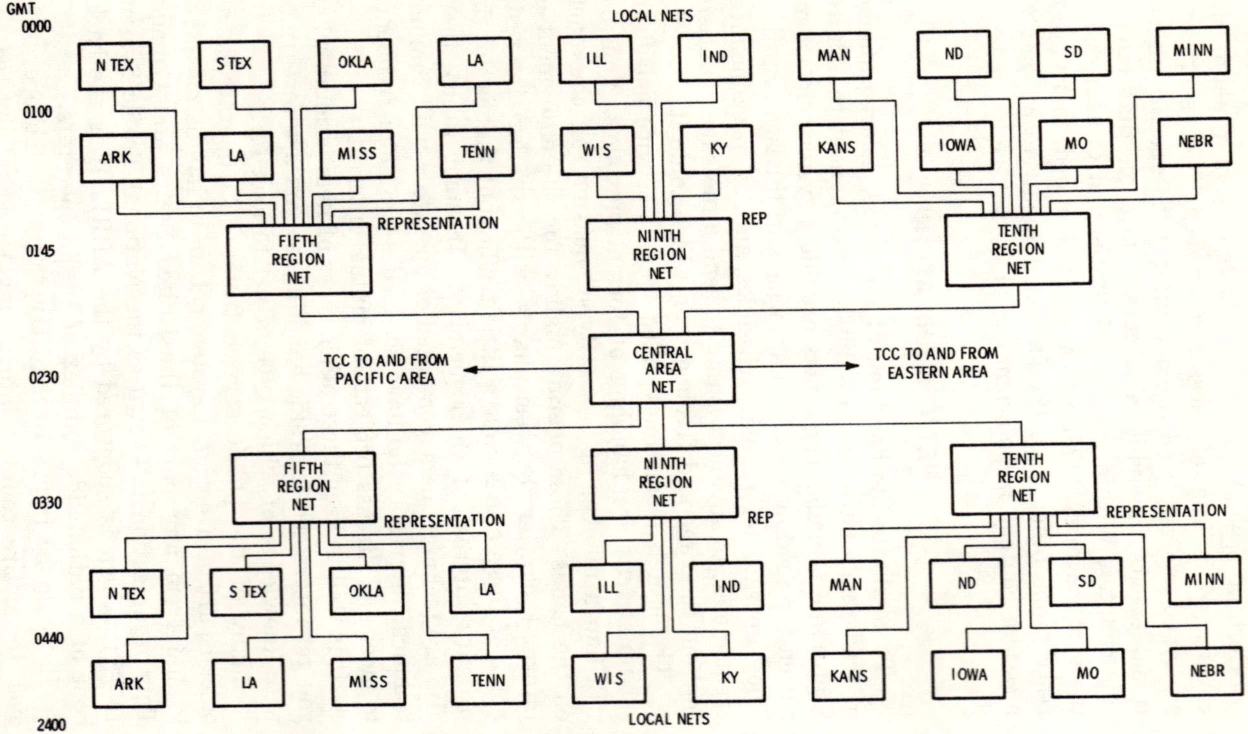
GMT
0000
0100
0145
0230
0330
0440
2400
LOCAL NETS
N TEX
S TEX
OKLA
LA
ILL
IND
MAN
ND
SD
MINN
ARK
LA
MISS
TENN
WIS
KY
KANS
IOWA
MO
NEBR
REPRESENTATION
REP
FIFTH REGION NET
NINTH REGION NET
TENTH REGION NET
TCC TO AND FROM PACIFIC AREA
CENTRAL AREA NET
TCC TO AND FROM EASTERN AREA
FIFTH REGION NET
NINTH REGION NET
TENTH REGION NET
REPRESENTATION
REP
REPRESENTATION
N TEX
S TEX
OKLA
LA
ILL
IND
MAN
ND
SD
MINN
ARK
LA
MISS
TENN
WIS
KY
KANS
IOWA
MO
NEBR
LOCAL NETS

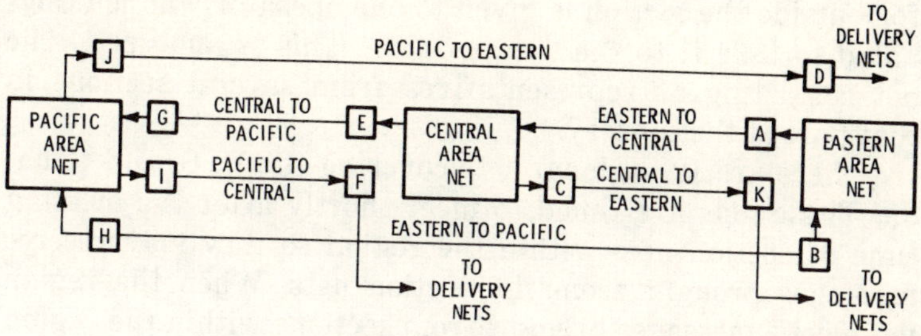

Block diagram at left illustrates interlinking of local, section, region and area nets in National Traffic System. Group shown here covers central area of United States and Canada; similar groups cover western and eastern parts of the nation.

Diagram above illustrates interlinking of the three area nets by means of operators taking part in Transcontinental Corps (TCC). The TCC operators do not form a net, but act individually on prearranged assignment to move traffic between the three area nets. Letters such as J, D, G, E, etc., represent these individual operators. The letters designate the title of their function within the system; that is, the J function indicates the PAN member designated to take messages going to the eastern area, and the G function indicates the PAN member designated to receive messages from CAN.

country is broken down into smaller areas, each having a net within it. Then, provision is made for selected representatives of these nets to meet at prearranged times to exchange messages from their respective areas. This arrangement keeps each net at a manageable size, and yet enables messages to be moved long distances in relatively short time.

The basic building block of the National Traffic System is the "section net." As previously stated in Chapter 4, the ARRL field organization is divided into "sections." In many cases, a state comprises a section, while in more densely-populated areas, a state may be divided into two or more sections.

Nets in these sections, composed of amateur operators in the section interested in traffic handling, meet each evening. Any amateur having traffic to send may check in. The traffic intended for locations within the section is transmitted to stations closest to the delivery address. Traffic for outside the section is given to one operator who is designated to take it to the "region net." This is another traffic net consisting of representatives from several sections in that part of the country.

There are 12 regional nets covering the U. S. and Canada. These nets are timed to meet shortly after the meeting time of section nets within the region so they may receive messages brought from the section nets. When the region net meets, messages brought from sections within the region are exchanged with representatives from each section, and any traffic going outside the region is given to an operator designated to meet the "area net."

Area nets are the "super nets" of the NTS—there are three area nets, which taken together span the entire United States and Canada. They are the Pacific Area Net (PAN), Central Area Net (CAN), and Eastern Area Net (EAN). Each area net is composed of representatives from region nets within the area. Traffic from a region destined for a location outside of that region is taken to the area net and given to another representative from the area for which the traffic is destined. Checking into each area net ordinarily will be two stations from each region within that area. One will bring traffic *from* the region net, while the other will receive other traffic to be taken *to* the region net.

After the area net members exchange their traffic, the representatives receiving traffic return to a second session of their region net, generally held a couple hours after the first session. At this second region net session, they give this traffic to operators representating the various sections within the region. These section representatives, in turn, take the traffic back to a second session of their section net later that night if there is one, or to the section net the following day.

This net organization is illustrated in the accompanying block diagram. It may appear complicated at first glance, but with operating times scheduled to interlink the various net levels, and with experienced operators handling the traffic, a large number of messages can be moved quickly and smoothly in an evening's time.

In addition to the ARRL National Traffic System, there are several other net organizations stretching across the nation. Among them are various Post Office nets and MARS (Military Affiliate Radio System) nets. There are separate MARS organizations for Army, Navy and Air Force. They operate on military frequencies rather than amateur frequencies, and include representatives from most military bases, as well as volunteer amateurs serving in the program. MARS traffic may move across the U. S. and also to military posts in some foreign lands where regular amateur third party traffic is not allowed.

Often messages may move part way to their destination on one net system, such as NTS, then be put into another system, such as MARS, at the discretion of an operator relaying the message. When messages are transferred from amateur to MARS channels, or vice versa, it is necessary to change the format of the message (but *never* the text!) to conform to the message format used by the amateur or military system into which the message is being placed.

In addition to the nets belonging officially to the National Traffic System, there are quite a number of "independent" nets of various flavors and sizes. Some cover an entire state, or maybe a couple of states, and some cover smaller areas within a state. Frequently these get started by a group of operators who fall into the habit of meeting at a certain time each day or each week to visit. Gradually, over many months or years, the group begins to handle a few messages in addition to visiting.

Often messages within the area covered by such nets can be relayed within the net itself to the destination. As the net grows in size and activity, generally there will be one or more members checking into it who also check into other nets, such as NTS, Post Office nets, MARS, or others, and so they can handle traffic going outside the immediate net area and bring back incoming messages picked up on other nets. Even though such nets are not officially a part of a nationwide traffic-handling organization, they still often form an important part of the overall capability of amateur radio operators to handle traffic.

In many parts of the country, there have been such nets formed to be available in case of a local emergency. Frequently such nets cover a city, or metropolitan area, or perhaps a county or two. Many of these have been formed by members of the Amateur Radio Emergency Corps (AREC). This organization is sponsored by the ARRL to provide a reliable means for amateurs to provide emergency communications if needed in their local communities.

AREC groups (see Chapter 4), along with the NTS membership, are grouped together in the Amateur Radio Public Service Corps (ARPSC). The objective of this overall grouping is to unite the local emergency net personnel with the nationwide traffic-handling capability of the NTS.

Often the local nets will have a weekly or monthly drill session, to maintain preparedness for an unexpected emergency, but in the meantime will participate in routine traffic handling. Many of these nets operate on six and two meters and have among their members a large number of Technician-class operators. These operators are not allowed to operate on the "low bands" where most NTS operating is done, but they can be an important part of the overall system by handling both incoming and outgoing messages within their local community.

NET OPERATION

Sometimes amateurs who have never participated in a net may feel timid or unsure of themselves and afraid to mingle with the "big boys" on the nets. There is no need for this, however, for one of the functions of nets is to provide training and experience. Every "hot shot" traffic operator was a beginner at one time, and got to his exalted position only through practice. It would be a mistake for the beginner to check into one of the very busy nets, such as an area net or busy regional net, but plenty of good experience may be obtained in local and section nets. After a good dose of this, the beginner will find traffic operators are eager to help him as long as he shows a sincere desire to learn. Before long, he will have acquired the experience to feel right at home with the "big boys" whom he used to look up at in awe.

Regardless of the size of a public service net, the most important ingredient it must have is *discipline.* Without this, the net soon breaks down into chaos, and little can be accomplished. To maintain discipline and keep everyone headed in the right direction, so to speak, each net session has a *Net Control Station* (NCS). This operator functions just like the control tower operator at an airport. He is in command of everything that is done on the net, and all net members must recognize that he is the undisputed boss so the net can be kept running smoothly.

On cw nets, the NCS generally does not use a roll call, but rather announces the beginning of the net and then stands by for stations to check in, or "QNI." Some phone nets use roll calls, and some do not. A roll call can get the net started a little smoother sometimes, but it also prolongs the net session greatly, and also may indirectly encourage members checking in to dilly-dally with excess comments and greetings that can slow down a net.

On cw, most experienced traffic operators use break-in, so the NCS can quickly hear a station asking to be recognized, and members can detect if they accidentally transmit at the same time as another station trying to check in.

Another important advantage to using break-in when sending traffic on cw is that if the receiving operator misses a word, he can close his key or send a string of dits to "break" the sending station and the sending operator will hear this and stop to be told what was missed. This can be done faster than waiting until the sending station has completed the message, and then give "fills" requested by the receiving operator. This same advantage is available, but to a somewhat lesser degree, with ssb when using VOX, since the sending operator will not hear the receiving operator trying to break him unless he pauses long enough for his VOX relay to open.

After stations have checked in and listed their traffic, or advised that they have no traffic, the NCS designates stations with traffic to call certain others who will receive it.

On cw, these stations usually are directed to move off frequency to move their traffic, then return to the net frequency to await further instructions from the NCS. This permits several messages to be moved simultaneously by having pairs of operators operating on several frequencies, usually spaced about 5 kHz or so apart. On phone, this also may be done if there are open frequencies available, but often traffic on phone nets must be moved on the net frequency, which means only one message may be sent at a time. In either cw or phone, the speed with which the messages may be moved is not limited by the mode of transmission, but rather by the speed at which the receiving station can copy the message.

MESSAGE FORMAT

To make message handling as systematic as possible, a standard format has been adopted by the ARRL and is universally used in the United States by amateur radio operators. Each of the MARS organizations have their own message formats which differ in details from the standard amateur format, because each is related to the military message form of the branch of service sponsoring the particular MARS organization.

There are four basic parts to the ARRL message format: the preamble, address, text and signature. The preamble is similar to the heading at the top of a business letter. It contains the message number, the "precedence designator," handling instructions, call letters of the originating station, the count of words and punctuation marks in the text, the place of origin, and the date of origin. Messages may also carry a filing time, but this is optional and often omitted unless it has particular significance.

Now, what do all these items in the preamble mean? The message number is a file number assigned to each message by the originating station. Most traffic operators start each year with "1," but the method is up to the individual operator. By having a number assigned to each message, it is easy to refer back to the message if necessary, for ex-

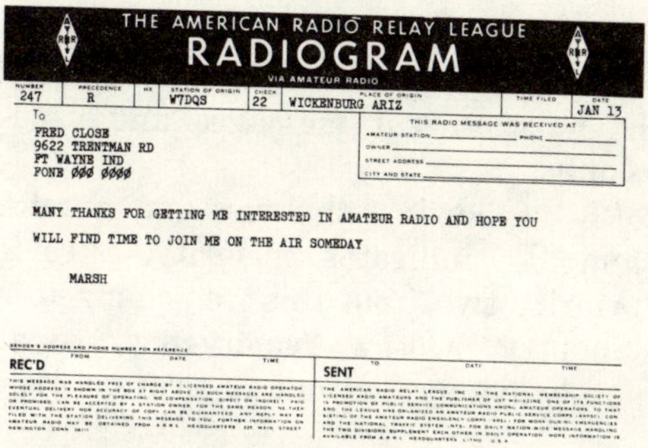

THE AMERICAN RADIO RELAY LEAGUE
RADIOGRAM
VIA AMATEUR RADIO

NUMBER	PRECEDENCE	HX	STATION OF ORIGIN	CHECK	PLACE OF ORIGIN	TIME FILED	DATE
247	R		W7DQS	22	WICKENBURG ARIZ		JAN 13

To
FRED CLOSE
9622 TRENTMAN RD
FT WAYNE IND
FONE ØØØ ØØØØ

THIS RADIO MESSAGE WAS RECEIVED AT
AMATEUR STATION ______ PHONE ______
OWNER ______
STREET ADDRESS ______
CITY AND STATE ______

MANY THANKS FOR GETTING ME INTERESTED IN AMATEUR RADIO AND HOPE YOU
WILL FIND TIME TO JOIN ME ON THE AIR SOMEDAY

MARSH

SENDER'S ADDRESS AND PHONE NUMBER FOR REFERENCE

REC'D	FROM	DATE	TIME	SENT	TO	DATE	TIME

Message blanks such as this provide neat, competent appearance for messages mailed or delivered in person to nonamateurs. Blanks also may be used for amateur files, although plain paper is acceptable.

ample if a station in the addressee's city is unable to deliver the message and sends a message back asking for more information on the address. (Such a message, relating to the handling of another message, is called a "service message," which is sent "SVC" on cw.) All messages handled by an amateur must be kept on file for a year, so by having a number assigned to each, any of them can be found in the file if needed.

The "precedence designator" follows the message number, and indicates the relative importance of the message. Normally, all message are "routine." However, a message having greater importance may be labeled "priority." This is used to indicate important messages having a specific time limit, and includes official messages not covered by the "emergency" category, and messages such as notification of death or injury, whether official or personal. In a net operation, priority messages are handled before routine messages, for obvious reasons.

Messages of inquiry on the health or welfare of someone in a disaster area carry the precedence "inquiry," and are handled after "priority" messages, but before "routine" messages.

"Emergency" is the precedence applied to any messages having life and death urgency, and includes official messages of welfare and rescue agencies during emergencies requesting supplies, materials, or instructions vital to relief of stricken persons in a disaster area. It is reserved for only the most urgent messages and used in serious emergencies only.

On cw, "R" following the message number indicates "routine," and "P" indicates "priority." Previously, "P2" indicated "priority two," but this was changed later to "Q," and called "inquiry." On cw, "emergency," as a precedence designator is always spelled out to prevent its being overlooked.

The next item in the preamble, the "handling instructions," is an optional item and is included only when necessary to comply with instructions from the person who gave the message to the originating station for transmission. A set of abbreviations, such as "HXA," "HXB," etc., are used to give instructions, such as "collect telephone delivery authorized," or "cancel message if not delivered within . . . hours," etc. The full meanings of these instructions are included in the accompanying box. These are used on a small portion of messages handled by amateurs, but are useful for certain special situations because they carry important information in a very short portion of the message.

The "station of origin" is the call sign of the station which first put the message on the air. This is the station which assigns the message number.

The "check" is a number indicating the total number of separate words and punctuation marks in the text of the message. By counting the words and punctuation marks in a message and referring to the check, an operator handling the message may get a clue if an error has occurred in transmission or receipt of the text.

The "place of origin" is the town or other appropriate location where the message originated. Ordinarily, this will be the town in which the originating operator lives, or when operating away from home, such as on a trip, it would be his location at the time he originated the message. If a third party in a different location gives the operator the message to transmit, then the "place of origin" would be that other location. This could happen, for example, if a person living in the city of Podunk called up a ham friend living in Centerville and asked him to send a message. Although the ham lives in Centerville, the actual place of origin in this case would be Podunk.

Emergency—Any message having life and death urgency to any person or group of persons, which is transmitted by amateur radio in the absence of regular commercial facilities. This includes official messages of welfare agencies during emergencies requesting supplies, materials or instructions vital to relief of stricken populace in emergency areas. During normal times, it will be *very rare*. On cw, this designation will *always* be spelled out. When in doubt, *do not* use it.

Priority—Important messages having a specific time limit. Official messages not covered in the "emergency" category. Press dispatches and other emergency-related traffic not of the utmost urgency. Notification of death or injury in a disaster area, personal or official. Use abbreviation "P" on cw. Inquiries as to the health or welfare of someone in the disaster area are handled after the above are cleared and are designated "inquiry" ("Q").

Routine—Most traffic in normal times will bear this designation. In disaster situation, traffic labeled "routine" ("R" on cw) should be handled *last*, or not at all when circuits are busy with emergency or priority traffic. Most traffic handled on amateur circuits in normal times will fall in this category.

The precedence follows the message number. For example, on cw, "207 R" or "207 EMERGENCY." On phone, "TWO ZERO SEVEN, ROUTINE (or EMERGENCY)."

Message Handling Instructions

HXA—(Followed by number) Collect landline delivery authorized by addressee within . . . miles. (If no number, authorization is unlimited.)

HXB—(Followed by number) Cancel message if not delivered within . . . hours of filing time; service originating station (send message to originating station advising the message was cancelled because expiration of the time limit, rather than delivered).

HXC—Report date and time of delivery (TOD) to originating station.

HXD—Report to originating station the identity of station from which received, plus date and time. Report identity of station to which relayed, plus date and time, or if delivered, report date, time and method of delivery.

HXE—Delivering station get reply from addressee, originate message back.

HXF—(Followed by number) Hold delivery until . . . (date).

A prosign (if used) selected from this list will be inserted in the message preamble before the station of origin. Example: "NR 207 R HXA50 W1AW 12 . . . . " If more than one HX prosign is used, they can be combined if no numbers are to be inserted, otherwise the HX should be repeated. Example: "NR 207 R HXAC W1AW . . . . " or "NR 207 R HXA 50 HXC W1AW . . . . " On phone, use phonetics for the letter or letters following the "HX."

The "time filed" is the time the message was composed by the originating operator, or the time it was given to him by a third party. This may be omitted unless it serves a significant purpose. "Date filed" is the month and day, but not the year, when the message was written down for transmission. This is usually, but not necessarily, the date it is first put on the air.

The "address," "text," and "signature" are all obvious, but certain precautions should be kept in mind. First, the address should be complete enough so the operator at the destination city who has to deliver the message can locate the addressee satisfactorily. Normally, the full name, street address, city, state, and, if possible, telephone number should always be included in the address. The delivering operator will normally try to deliver the message by telephone if the addressee lives in the same community, but if no ham can be located on a net in the destination city, an amateur operator in some other part of the destination state may need to mail the message to the addressee. This means the address should include the proper information so the message may be delivered by mail or by telephone, or so anyone can locate the addressee by means of a telephone directory or assistance from a telephone operator.

It is generally urged that the text be limited to 25 or 30 words. By eliminating all unnecessary words and writing the message in the brief, punchy language that is typical of telegram style, it is possible to get quite a bit of information into 25 words.

SENDING A MESSAGE

Transmitting a message is very similar to transmitting anything else that is written down in front of you, but there are a few special items included to be sure the receiving station gets the message written down in the same format as you have it.

On cw, for example, "$\overline{\text{BT}}$" is sent following the preamble, the address, and the text to break up the four main parts of the message. "AA" is sent at the end of each line in the address to indicate the end of the line. This prosign may be used in the signature if needed if there is a title or address to go with the name in the signature and it is desired that these appear beneath the name when the message is written down for delivery.

Punctuation is *not used* in the address and is used as little as possible in the text. In composing a message to be put on the air, it is helpful to omit punctuation if possible, especially if accidental loss of any punctuation could cause the message to be misunderstood. On cw, where everything is spelled out and there are code characters to indicate punctuation marks, any punctuation should survive radio transmission. However, on phone, an operator reading the message into a mike may neglect to specify punctuation marks in the text, because we are accustomed in reading aloud to use our voice inflection to indicate punctuation, rather than call out the punctuation marks. This long-standing habit is often carried over subconsciously when speaking via radio. This example indicates the great attention that must be given to small details when sending messages by voice. The human mouth is actually the most unreliable vehicle of communication that exists, so a competent voice operator takes great pains to be sure he is properly understood.

On cw, some operators use "X" in the text to indicate a period, while other operators spell it out or abbreviate its spelling, like this: "PERIOD" or "PD." Phone operators sometimes use the phonetic "X-RAY" to indicate a period, although saying "PERIOD" is generally preferred.

On cw, a comma may be sent with the comma symbol or spelled out or abbreviated like this "COMMA" or "CMA." However, it is standard to indicate a question mark in the text with the word "QUERY." This is the accepted practice on both cw and phone.

It is important to *never* use a question mark in the text of a message when sending it on cw because of the likelihood of a misunderstanding. This could happen because cw operators sometimes send the question mark after they make an error in transmission, then resend the word correctly. Also, a question mark is used when the receiving operator asks for a "fill" of a part of the text he missed. If there was actually a question mark in the text, it could be confused with these other standard uses of a question mark in cw operating. Consequently, "QUERY," which is always spelled out, is the correct way to indicate presence of a question mark in a message sent on cw and is generally accepted on phone operations as well. In all cases of punctuation in the text, whether the actual punctuation mark is sent, or spelled out or abbreviated, each punctuation item counts as a word in the "check" in the preamble.

After finishing the signature on cw, "$\overline{\text{AR}}$" is sent, followed by "N" if the sending operator has no more messages to send to the receiving operator, or "B" if he does have more messages to send to him.

On phone, just saying "end of message" is satisfactory, and can be followed with something like "two more to follow" or "no more to follow" if desired.

Sending a message on phone is pretty much a matter of reading the message clearly, and doing it slowly enough for the receiving operator to write it down. Any names or other words which might be misunderstood should be spelled phonetically. Following the preamble, say "to" before giving the address, and say "break" following the address before starting the text. "Break" is used again at the end of the text before giving the signature.

Some phone operators will stop while transmitting a message and say "break for check," then pause to see if the receiving operator is copying satisfactorily. This was more commonly done when a-m was more widely used, since the transmitter carrier normally stayed on until it was switched off by the operator. The use of VOX with ssb makes this less necessary, although it can still be helpful, especially in long transmissions.

If you have cw break-in capability, it is easy for a station to break you for fills as you go through the message, rather than take extra time afterwards to ask for them. Break-in capability is a tremendous advantage in cw traffic handling. On ssb, a similar advantage is available using VOX if you pause often to allow the VOX relay to open.

A few brief prosigns are used on cw to speed up the procedure of asking for fills. Included are "WA" meaning "word after," "WB" meaning "word before," "AA" meaning "all after" and "AB" meaning "all before." To get a fill of material missed between two words, send the last word you got correctly, then a question mark, then the next word you got correctly. In giving you a fill, the sending operator generally will resend the last word you got, then the fill, then the word following the fill, so you can verify that the fill you received fits into the message.

For example, if a part of the text reads "BEST WISHES ON YOUR BIRTHDAY AND HOPE TO VISIT YOU NEXT MONTH" and you missed the words "BIRTHDAY AND HOPE," you would send this to get them: "FILL YOUR ? TO," or "BN YOUR ? TO." ("BN" means "between.") The sending station would reply this way: "YOUR BIRTHDAY AND HOPE TO." The real traffic hot shots deplore the sending of the words just before and just after the actual fill material, because they feel this is needless waste of time and are insulted when someone does this to them. However, for the beginner, it is generally good to send these extra words to be sure the receiving station gets the material correct.

If you are not sure you copied something correctly, send "CFM" (meaning "confirm") followed by the material of which you are not sure, followed by a question mark. If you have the material correct, the sending operator will reply "CFM" (meaning "confirmed.") If you got it wrong, he will send "N" (meaning "no"), then resend the material you questioned. On phone, it is done pretty much the same way, except the request for missing material and the responses are done orally.

On rtty, the procedure follows that used on cw except for a few special items: (1) extra spaces are used between lines of the address instead of "AA"; (2) line spaces are used to separate text from address and signature; (3) a "CFM" line is sent after the signature, repeating all names, numerals and unusual words in the message in the order originally transmitted.

To illustrate the sending of a complete message by cw, phone and rtty, let's use the message shown on page 101, where it is shown on an ARRL message delivery blank as it might be mailed or personally given to the addressee.

When sent on cw, this message would be transmitted this way: "NR 247 R W7DQS CK22 WICKENBURG ARIZ JAN 13 $\overline{\text{BT}}$ FRED CLOSE AA 9622 TRENTMAN RD AA FT WAYNE IND AA FONE 000 0000 $\overline{\text{BT}}$ MANY THANKS FOR GETTING ME INTERESTED IN AMATEUR RADIO AND HOPE YOU WILL FIND TIME TO JOIN ME ON THE AIR SOMEDAY $\overline{\text{BT}}$ MARSH $\overline{\text{AR}}$ N."

On phone it would be read this way "NUMBER TWO FOUR SEVEN ROUTINE FROM W7 DAVID QUEEN SUSAN CHECK TWO TWO WICKENBURG ARIZONA JANUARY ONE THREE TO FRED CLOSE I SPELL THE LAST NAME CHARLIE LEWIS OTTO SUSAN EDWARD NINE SIX TWO TWO TRENTMAN I SPELL THOMAS ROBERT EDWARD NANCY THOMAS MARY ADAM NANCY ROAD FORT WAYNE INDIANA TELEPHONE ZERO ZERO ZERO ZERO ZERO ZERO ZERO BREAK MANY THANKS FOR GETTING ME INTERESTED IN AMATEUR RADIO AND HOPE YOU WILL FIND TIME TO JOIN ME ON THE AIR SOMEDAY BREAK MARSH I SPELL MARY ADAM ROBERT SUSAN HENRY END OF MESSAGE."

On rtty, it would be sent this way:

```
NR 247 R W7DQS CK22 WICKENBURG ARIZ JAN 13
FRED CLOSE  9622 TRENTMAN RD  FT WAYNE IND  FONE 000 0000

MANY THANKS FOR GETTING ME    INTERESTED IN AMATEUR RADIO AND
HOPE YOU WILL FIND TIME    TO JOIN ME ON THE    AIR SOMEDAY

MARSH
CFM CLOSE 9622 TRENTMAN 000 0000
```

Sample message in format used for sending via RTTY. Text words are spaced in five-word groups to make it easier to check the count on receipt of message over the air. Confirmation line at the bottom is a feature unique to RTTY messages and includes names, numbers, and any unusual spellings of words in the message as a double check against garble in transmission that might otherwise go unnoticed. Some operators on cw and phone train themselves to copy text in five-word groups as shown here for convenience in counting words quickly after receiving the complete message.

TRAFFIC-HANDLING SHORTHAND

A list of "standard texts" has evolved down through the years for use by traffic handlers to speed the movement and delivery of many messages which fall into certain broad categories. By having a list of the most often used types of such messages with a number assigned to each one, it is possible for traffic operators to move such messages more quickly, since they can transmit the number rather than the full text.

These messages are called *ARRL Numbered Radiograms.* A list of them is included in ARRL log books, and the messages are referred to as "ARL texts" on the air. When used, each number is preceded by the letters "ARL" to designate that the number following refers to a standard ARL text. On cw, the numbers of the texts are always spelled out to avoid error.

For example, message twenty one means "YOUR GIFT GREATLY APPRECIATED. MANY THANKS." This text would be sent "ARL TWENTY ONE." Nothing more is needed. Before delivering such a message, however, it is vital that the amateur delivering the message to the nonamateur to whom it is addressed must look up the meaning in his ARL message list and replace the number with the text meaning, since the number obviously will not mean a thing to a nonamateur.

ARL message texts may be combined with other words in a message or they may be used alone as the complete text. More than one ARL number may be used in a message, if desired. In any case, the "check" in the preamble of the message contains the letters "ARL" followed by a count of the words and numbers in the text—including the spelled-out numbers of the ARL text number. For example, if the text as transmitted reads "ARL THIRTY ONE HOPE TO VISIT YOU NEXT SUMMER," the check for this text would be ARL 9, to show that the text includes one or more ARL texts, and that there are nine words in the *transmitted text,* although the full text as it will be delivered to the addressee, after "decoding" of the ARL message number, will be greater than 9. A list of these ARL messages appears in the appendix.

SPECIAL PROWORDS

In Chapter 1, explanation was given of special uses of "break" as a proword by the WCARS-MWARS-ECARS nets on 40 meters. The full list of prowords used by these nets are:

Break Break Break—Emergency only, all stations stand by. Woe betide the unwary operator using the triple break for casual interruption.

Break Break—Priority or urgent traffic.

Contact—Used to notify the NCS that you wish to contact a station just heard.

Information—Used to notify the NCS that you have information pertinent to the business at hand.

Recheck—Used to notify the NCS that you have failed to make contact with a station just moved off the regular net frequency.

Question—Used to notify the NCS you have an inquiry regarding the business at hand.

The WCARS procedure manual cautions: *"Never say more than one brief sentence without dropping your VOX or mike button."*

DELIVERING MESSAGES

Delivery of messages to nonamateurs by the amateur operator who is nearest the destination generally is done by telephone, if possible. It is always wise to check the local telephone directory before calling the addressee to be sure the phone number in the message address is correct.

Amateurs are not expected to make long-distance calls at their own expense to deliver messages, but they may if they want to. If the addressee lives in an area that would require a toll call, the delivering operator may place a collect call to that person, and explain to the telephone operator that he is unknown to the party being called, but that he has a radio message for that party from a person by the name of such-and-such. The operator then can explain this to the party being called, since otherwise they would be unlikely to accept a collect call from a stranger. In cases of urgent messages, an amateur operator may decide to pay for the call himself, in a spirit of public service, because of the importance of the message. Often grateful persons receiving such messages will reimburse the amateur operator for his telephone expense.

Three special public service nets have been set up by dedicated amateurs on 40 meters to provide a full-time monitored net available to home stations as well as mobiles on the road. Here, members of West Coast Amateur Radio Service, which pioneered the 40-meter full-time net idea operate a portable station at an amateur radio meeting.

An alternate method of delivering urgent messages is to give a copy of the message to a police department near the amateur's home and ask if they will relay the message to the destination through their own communications channels.

If the destination is in a location away from the amateur's area and telephone delivery does not seem necessary, the message may be mailed by the amateur operator. Some operators question the need for spending postage money on behalf of a stranger, but it may be realized that most of us spend more on a minor accessory for our station than we would spend in several years of mailing a moderate number of messages. Since message handling is a public service which is a key function of amateur radio, it seems that spending a small amount for postage and envelopes could be looked upon in the same way as spending a much larger amount on a new antenna or a fancy mike.

For delivery by mail, or in person, attra
like message blanks are available from ARR
message on page 101 appears on such a message blank. Some nets print their own message blanks for use by members, and MARS stations have specific message blanks for this purpose. Whichever method is used—even a blank piece of paper can suffice—it should be kept neat and competent-appearing to give the addressee receiving it a favorable impression of the message service which has been provided.

EMERGENCY PREPAREDNESS EVENTS

Two key events are staged annually to test the ability of amateur operators to respond to emergency situations. One is the annual *Field Day* contest, and the other is the *Simulated Emergency Test* (SET).

The main object of Field Day is to test amateurs' ability to operate under emergency conditions, when commercial power might not be available. It takes place on a weekend in June, and is engaged in by thousands of amateurs, mostly members of radio clubs, who set up radio equipment outdoors or in locations not normally used for their day-to-day radio operations. Gasoline generators are used for most operations, and antennas generally are set up in locations away from existing buildings.

The whole operation takes on the aspect of a radio picnic, but the objective is a serious one—practicing the capability to stay on the air in case of a large-scale emergency.

During the *Simulated Emergency Test,* which occurs on a weekend in January, traffic operators enlarge the number of net sessions and originate a much larger number of messages than ordinarily handled in routine times. In this case, the objective is to shift the nation-wide traffic system into high gear to test everyone's ability to handle a large number of messages. Many amateur emergency communications leaders also stage local simulated emergencies, involving mobile and portable stations, to see how well their emergency amateur workers can operate with sudden unexpected difficulties thrown at them.

RACES

A long-standing amateur emergency organization which you occasionally hear about is *RACES*—Radio Amateur Civil Emergency Service. The prominence of RACES activity will vary greatly from one part of the country to another. In some places, RACES is very active, while in other places, it is virtually unknown. This wide variation usually is caused by the varying need for such a service in different areas, and by the varying interest shown in it by amateurs in these areas.

RACES is the Civil Defense arm of amateur radio, with special FCC regulations governing it. The objective has been that in case of a large-scale disaster situation in the United States, regular amateur radio operations would be prohibited. Theoretically, the nation's Civil Defense organization would swing into action, and stations involved in RACES operations would be allowed to continue operation, under supervision of Civil Defense officials.

For this reason, areas with active RACES organizations have "RACES plans" drawn up for government agency approval. These plans provide for operation on specific frequencies by amateurs enrolled in RACES, even if other amateurs were ordered off the air by the government.

The program originated following World War II and is based on a premise of that time that radio stations in the United States could provide "homing signals" for enemy military weapons. Consequently, it was felt that normal radio operations should be suspended to eliminate such signals, but that provision should be made for certain radio operations to be authorized for necessary emergency services. Modern technological developments which provide much more sophisticated guidance for enemy military forces seem to make this theory quite obsolete.

OVERSEAS TRAFFIC

Amateur traffic nets normally cover only the United States and Canada, but third-party traffic can be sent to certain overseas points by amateurs and to U. S. military personnel in additional ones through the MARS military radio circuits, to which amateurs have access through their local area MARS nets.

Volunteer amateur members of the *Military Affiliate Radio System* have handled thousands of phone patches for servicemen by linking their homes in the United States with the servicemen overseas. An outstanding leader in this activity has been Barry Goldwater, KU7GA, whose MARS station, AFA7UGA, is at his home in Phoenix, Arizona. Assisting him are about 25 Phoenix amateurs who keep the station operating at all times during which propagation conditions allow contacts with the overseas bases.

Some countries specifically prohibit the handling of third-party traffic with other countries by means of amateur radio. As this book is written, the latest available list of countries to which amateur operators in the United States may send third-party messages via hams in those countries is: Argentina, Barbados (only U. S. stations/8P), Bolivia, Brazil, Canada, Chile, Colombia, Costa Rica, Cuba, Dominican Republic, Ecuador, El Salvador, Greenland (XP calls only), Haiti, Honduras, Israel, Liberia, Mexico, Nicaragua, Panama, Paraguay, Peru, Uruguay and Venezuela.

The permissible call sign prefixes in these countries are: CE, CM, CO, CP, CX, EL, HC, HH, HI, HK, HP, HR, LU, OA, PY, TI, VE, VO, W or K/8P, XE, XP, YN, YS, YV, ZP, 4X, and 4Z.

Amateur operators in Canada may send third-party messages to amateurs in Bolivia, Chile, Costa Rica, Dominican Republic, El Salvador, Honduras, Israel, Mexico, Peru, U. S., and Venezuela. The permissible prefixes in these countries are CE, CP, HI, HR, K, OA, TI, W, XE, YS, YV, and 4Z.

Note that these lists do not refer to the countries with which amateurs may make contact; these lists are strictly for handling formal messages on behalf of nonhams. Amateurs are allowed to conduct informal "ham-to-ham" visits on the air with nearly all countries in the world—the exceptions are detailed in Chapter 10.

Messages to or from United States military personnel stationed at foreign bases may be filed with MARS stations, either those on military bases, or those operated in their homes by amateur MARS members. Traffic sent to overseas military personnel and injected into the amateur National Traffic System is sent to a stateside MARS outlet which then places it in the MARS overseas system.

Occasionally special permission is granted by foreign countries for exchange of third-party traffic with the United States through regular amateur channels when it is related to certain special events. For example, some foreign countries in southeast Asia, where there have been many United States military personnel stationed in past years, have approved amateur radio contacts at Christmas time to enable the handling of Christmas messages between U.S. servicemen and their families.

MARS stations have been especially active handling phone patches for servicemen stationed overseas. These calls are handled direct between the United States and foreign military bases by MARS operators on both ends. They are patched into the United States telephone system by the MARS operator in the United States, who places a collect telephone call to the family of the serviceman. Operators handling these calls report the way to arrange for such a contact is to write to the serviceman overseas and ask him to contact the MARS station on his base and request a call, which is originated from that foreign base, *not* from a MARS operator in the United States.

Another important type of traffic for which amateurs often receive requests is the notification of a serviceman overseas of a death or serious illness in his family. Persons experienced in this type of message recommend that the family contact its local Red Cross representative, who will arrange for the message to be sent to the serviceman through Red Cross channels. The Red Cross also can arrange for any necessary verification of the family death or illness to be sent to his commanding officer if it may be necessary to request a leave for the serviceman to return home.

SECRECY AND RECORDS

Amateurs are required by FCC rules to keep a copy of each piece of formal traffic they handle on file for one year. This is not just messages originated by the station, but messages received or relayed as well. It is also important to remember that all such messages come under the federal "secrecy of communications" regulations, which provide that their content may not be divulged to anyone except the addressee, to someone representing him, or to another station operator to whom the message is being sent.

In other words, either on the air in casual conversation or off the air, idle gossip regarding messages a ham has handled, is prohibited. This may seem unusually strict to some people who look upon ham radio as consisting entirely of informal and casual visiting, but it should be remembered that amateur operating is not all fun and games—it has its serious side as well, and including ham radio in the secrecy of communications regulations emphasizes that it is considered a serious activity by the men who make the rules we live by.

Chapter 8

Contest Operating and Certificate Chasing

Among the most fervent participants in any amateur radio operating activity are the contest enthusiasts. When a beginning ham tunes across a band during a contest weekend, he generally gets the feeling that the world has gone mad, and that all the hams on the globe are transmitting simultaneously and are screaming hysterically at each other. However, with a little more experience, the casual observer begins to sort out the sense from the noise, although he is still amazed that anyone can really keep track of what is going on, since the chaos level seems to be many miles deep.

The very fact that participants in a contest *can* keep track of what is going on points up a key purpose of contests—to sharpen operating skills. Since an unusually large number of operators are on the air during a contest, it is necessary for anyone compiling even a moderate score to be fairly skillful in his operating techniques. If he isn't, he won't make many contacts.

Jumping into a contest the first time can be a terrifying experience for someone accustomed to the slow, easy-going pace of an informal hour-long visit on the air with a good friend. At first, it seems the whole world is assaulting your ears with whoops and howls. Gradually, though, an operator learns to distinguish between some of the whoops and howls—this marks the first step in the sharpening of the ears of a beginning contest operator.

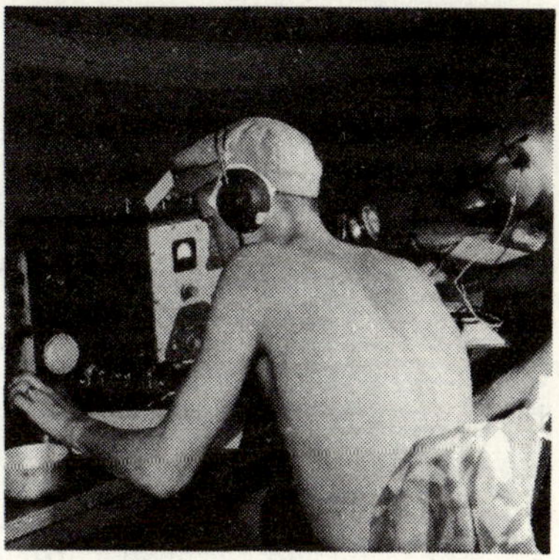

One of the most popular contests each year is Field Day, which tests both general operating skill and the ability to set up a station away from home without use of commercial power. This team of operators is set up in a tent in a farmer's pasture field. One operator runs the rig and makes the contacts, while the other fills in the log book and check-off sheet.

A successful contestant trains himself to mentally concentrate on a single signal and ignore the many other signals that are very near to the frequency of the signal he wants to hear. Some active traffic handlers say that contest operating is good training for a traffic station operator and especially for a net control station operator, for it trains him to listen carefully and selectively. On cw, it also trains an operator to copy code well under chaotic conditions; on phone, it trains an operator to speak clearly and briskly and to keep his transmissions short and to the point.

In a contest, stations making contact exchange certain information with each other, and this information becomes a part of the contest log, submitted to the sponsors for scoring. This information is known as the "exchange," and consists of call letters, plus such typical items as location, number of the contact, signal report, or other information pertinent to the particular contest.

Full break-in capability is a tremendous advantage in making these exchanges quickly, for getting fills of missed material and in hearing if any other station is calling at the same time you call.

A fringe benefit from contest operating can be the acquisition of numerous additional QSL cards, either to build up the prestige of the normal collection or to obtain some seldom-heard locations needed for a certificate such as Worked All States or DXCC. A contest brings a lot of operators out of the woodwork all at once, so there is an increased chance of obtaining a QSL from a location that has been missing from the card file.

The fast pace of a contest gives the feeling that it is a dog-eat-dog situation, with everyone trampling on everyone else's toes in a mad dash for a high score. Veteran contest operators do not feel this way, usually. They admit it is a mad scramble for points, but most of them take pride in being good sports about it. In other words, if they do stomp on someone's toes, everyone is supposed to understand that it is all in fun.

These battle-scarred veterans of many hours of grinding away during a heated contest point out certain unwritten rules of good sportsmanship that are adhered to by the sincere contest operators. For example, an operator frequently will build up a "string" of stations calling him, one after another, possibly because he is in a little heard location, or has an outstanding signal. It is considered very bad manners among contesters for one of these stations to take over that frequency and look for additional contacts before the first occupant has made contact with all the stations waiting to call him and has run the frequency "dry" for himself.

A matter of honesty in reporting the amount of power used when submitting the log for the contest also is considered important among rabid contest operators. Some contests have a "power multiplier" included in the scoring methods, so that lower-power stations, which may not be able to make contacts as easily as high-power stations, are given a handicap allowance to make their scores more nearly on a par with higher-power stations.

It is impossible for contest officials to come around and check everyone's power level, so everyone is put on the honor system. Anyone ever discovered cheating, or even suspected of it, will be black-listed by active contest operators in a very effective manner—they just won't work him on the air anymore!

As with all radio operating, it is important when using cw in a contest to send at the same speed as the station which called you. Electronic keyers have become popular with a number of contest participants because the sending speed can be quickly adjusted by turning a knob on the keyer. Some operators using bugs actually have two or three of them side by side, with the pendulum on each one set for a different speed, or they may keep a couple of extra weights handy on the operating table to be quickly dropped on the pendulum to slow it down. At the very least, an active contester will have a bug set at his favorite operating speed and also have a straight key for working other speeds.

This courtesy of adjusting sending speed to match the speed of the station calling is a mark of a gentlemanly contest operator who is interested in making contacts, rather than in putting on a speed demon act on the air. Adopting operating tools which enable such a fast change of sending speeds also is typical of the dedicated contest operator. Many such operating conveniences, all having as their objective a smooth and fast station operation, are used by contesters.

Arranging equipment so controls will be easy to reach, and having a writing area with log and check list in handy locations are typical of many refinements each operator works out to suit his own personal operating tastes and habits. Even the location chosen for the station clock can be important; putting it where the operator can glance up from the log without wasted time and effort can save seconds which are important in making more contacts in the time allowed for the contest.

Some operators build a part of their station with an eye to being able to switch bands quickly during a contest, and some will place a high premium on such items as the type of tuning knob or the placement on the panel of the rf gain control of a receiver. These seemingly minor items can greatly affect operating convenience in the pressure of a contest.

IMPROVING OPERATING EFFICIENCY

Using high power might at first seem the way to be head and shoulders above the competition in a contest, but experienced operators say this is not necessarily so. Although having a stronger signal provides an advantage in making contacts, the power level necessary to attain it may create a hardship in the final score because of the power multiplier used in some contests. Also, the additional tuning controls on a high-power final amplifier must be readjusted when switching bands, or when moving from one part of a band to another, and this consumes valuable time.

A good receiver and good antenna system, and the operating skill to make good use of them generally are more important than high power in contest operating. The type of antenna used may vary, depending on whether the contest is domestic or DX. Stateside contacts may be made best with an antenna having a higher radiation angle than that which may be preferred for DX work. The antenna that works best for a traffic station covering one state might be very poor for the contest operator using the same band to cover a continent.

Careful planning of operating times and bands, taking into consideration anticipated band conditions, is one of the many applications of operating skills that can help achieve a higher score in a contest. For example, 20 meters might be in great shape for stateside or even DX contacts in the daytime, but will go virtually dead at night during times of low sunspot activity. So the shrewd contest operator might operate on 20 during the day, then switch to 40 or 80 at night, when those bands will be in better shape for longer distances than they are in the daytime.

During sunspot peaks, 20, 15, and 10 will be good almost around the clock for long distances, but during sunspot minimums, they will be good only in the daytime, and sometimes not even then. However, 40 and 80 will be in good shape for stateside contacts in the daytime and will be usable at night for DX work during sunspot lulls.

So, the choice of band will vary, depending on the type of contest (domestic or DX), the time of day, time of year, and period in the sunspot cycle. Understanding how these factors affect radio propagation is one of the many ways in which skillful use of radio can have a direct bearing on a contest operator's success. Merely spending long periods of time frantically shouting into the mike or banging on the key are not enough to do the job.

Some operators adopt a regular pattern for scanning a band, systematically seeking contest contacts, and find it pays off better than random back and forth scanning. For example, an operator might pick a busy frequency and "work it dry," then move up a specific distance in the band—(maybe 5 or 10 kHz) and repeat his technique with a new group of operators which he usually will find on that frequency. By following a pattern that they have found to be effective, such operators can thoroughly cover a band, then start over, or switch to another band and repeat the system all over again.

DOUBLE-PURPOSE CONTESTS

One of the contests which attracts thousands of operators each year is Field Day. This differs from all other contests in that the main objective is to operate "in the field" using portable generators and temporary antennas. A purpose of this contest is to encourage operators to maintain equipment that can be moved into any location and set up for operation without dependence on commercial power.

Many Field Day operations are manned by members of a radio club, and several transmitters and antennas are set up at a single site, with each on a different band. In addition to enabling the group to achieve a higher score, this method also further simulates a disaster communications setup, in which more than one band probably would be used to maintain communication with different areas at different distances.

Some radio clubs have outfitted trailers, trucks, or buses for emergency communications work, and use these at Field Day time for their radio operations. Radio sites are usually in open fields or parks, where there will be no nearby residents to be bothered by the noise of round-the-clock operating with gasoline generators running continuously. Mountain tops in western areas are often used for the obvious advantages they provide, especially in vhf and uhf operating.

Often a radio club will use some of its funds to buy and maintain a generator for emergency communications and Field Day work, and club members will loan their home equipment when needed for such operations. This can provide maximum benefits in a small club that may not be able to afford to outfit a complete station for club use. It also makes emergency operation, without dependence on commercial power, available to club members without the individuals needing to own their own power generators.

One particular series of contests staged each year serves as good training for operators desiring to get their feet wet in contest operating without plunging into something as hectic as the annual sweepstakes or DX contests. This series is the "CD Parties" sponsored by the ARRL for all League officials and members of the field organization. "CD" indicates Communications Department—the department at ARRL headquarters, which supervises and coordinates the field organization appointments such as SCM, SEC, RM, PAM, ORS, OPS, OO, OVS, etc., as explained in Chapter 4.

When you hear someone calling "CQ CD," the purpose is for a contact in this contest. Normally, participation is limited to ARRL officials and holders of communications-department field appointments, but some parties have been "open" parties for any ARRL members to join in the activity. The pace of these contests is brisk, but still has an air of friendliness to it since everyone has a common interest besides that of the contest itself. By participating in one of the "open" contests, or by setting aside some contest operating time after acquiring an ARRL appointment, any ARRL member can get some practical contest operating experience before plunging into one of the massive contests open to all amateurs.

To encourage Novices to participate in contests, there is an annual Novice Roundup, which gives beginners an opportunity to learn contest operating by competing with others who have about the same level of experience.

Because of the large number of contacts that may be made in a short time during a contest, it is often possible to quickly build a large QSL collection based merely on contest contacts. In fact, it is not uncommon for a hustling operator to acquire a WAS (Worked All States) certificate solely on the basis of contacts made during one contest weekend.

CERTIFICATE CHASING

WAS and DXCC (DX Century Club—for 100 DX contacts) for many years have been two of the main operating certificate awards given to hams. As increasing numbers of operators have acquired these certificates, there has been a growing desire for a higher award to shoot for. Consequently, the "Five-Band" awards were created to provide an increased challenge for such operators who enjoy the pursuit of QSL cards from new locations.

These awards are given for achieving the award's goal fully *on each of five bands.* In other words, 5BDXCC (Five Band DX Century Club) is given for working 100 countries on each of five bands. For 5BWAS, the award is given for working all 50 states on each of five bands.

Certificates, sometimes called "wallpaper" by amateurs, are given as recognition for accomplishments in wide variety of amateur activities. These include traffic handling, working 100 or more foreign stations, working all states or all counties within a state, and even for rag chewing itself. Whatever the interest, there is a certificate for it.

Of course the "conventional" DXCC and WAS awards still are available, as are many others such as Worked All Continents, and a large number of certificates available from various localized radio groups for working all counties within a state, working a certain number of stations or counties within a state, and many others.

There are a large number of certificates and awards available from many national and local radio groups—so many that there are special clubs and newsletters devoted solely to this activity.

For the amateur interested in "certificate chasing," a "QSO file" is a handy station accessory. Most hams use either a 3 × 5 card file or one of the rotary-type files now commonly used in offices. A fresh card is filled out for each station worked, and includes data on location of the station, band on which the contact was made, whether a QSL card has been received verifying the contact, and the date of the contact. The latter makes it easy to find the contact in the log book if needed, and also shows at a glance whether the contact was made at a time when contacts for a particular certificate are recognized.

By having such a file available, an operator has a continuously updated check list with which he can quickly determine if he is eligible for any new certificates which he hears about, or he can determine what locations or other gaps still remain in his list of contacts before he can obtain the certificate. Some fellows practically cover the walls of their shacks with certificates of various kinds obtained through radio operating, which explains why radio certificates often are called "wallpaper."

Chapter 9

DX Operating

Frequently a nonamateur will ask of an amateur: "How far away can you talk on your radio?" The logical answer is: "Anywhere in the world."

This usually makes the questioner stop to ponder the thought, for it generally takes him by surprise, and may make him feel you are being a wise guy. However, it is obviously true that on the frequencies and with the power levels permitted for amateur work, it is possible to carry on two-way radio contacts with other stations anywhere in the world.

This type of operating, which has become known as "working DX," has become so popular that it is a whole separate phase of amateur radio. Most operators who really concentrate on DX do nothing else, for to do it well generally takes all the time that the average ham has available for operating.

DX is an abbreviation for "distance," and in the early days of radio, "working DX" meant making contact with other stations a considerable distance away, even though still in the same country. DX still has this meaning for vhf operators, who have only certain limited opportunities for overseas contacts. But for all other amateurs, DX has come to mean contact with foreign countries.

Not all DX work by hams has been done strictly for the satisfaction of being able to brag about how many foreign countries have been worked. Many times amateur operators have maintained the only communications links with various scientific expeditions, flights, and voyages in all parts of the world.

Successful DX operating requires skillful operating techniques, plus a convenient station layout. This station is equipped for operation on cw and ssb, and includes phone-patch facilities which have been used to provide voice contact for scientific and military personnel at the South Pole with their families in the United States. The globe serves as a beam-antenna direction indicator; a motor in the base turns the globe so a wire pointer indicates the area of the world toward which the beam is aimed.

This experience and capability has paid off handsomely for such extensive scientific programs as those established in Antarctica, where darkness prevails for half the year and few trips in or out can be made by personnel working there during this period. Through amateur radio, military support personnel and civilian scientific workers kept in touch with their families via ham radio—which gave a true humanitarian purpose to amateur radio DXing.

Other examples of DXing with a serious purpose include a number of nets and operating schedules maintained by stateside stations on behalf of church missionaries in foreign lands, many of them in primitive areas without other communications facilities. Also, hams often have put

DX capabilities to work to assist in medical emergencies in foreign lands by obtaining specialized medical information or rare medicines from stateside sources.

THE CHALLENGES OF DX

Many countries have a sizable number of hams, and making contact with at least one of them is not too hard. But small countries with few hams in residence (sometimes only one) are much sought after prizes, for which avid DXers will work months to get a 30-second contact.

The challenge and competition of working a large number of foreign countries and obtaining QSL cards verifying the contacts has spawned specialized operating techniques and their attendant skills. Anyone can work a few countries, but to work a lot of them takes certain qualities that make serious DXing a very specialized form of radio operating.

The qualities that go to make up a good DX operator are combinations of operating skills and special forms of courtesy that apply only to DX operating.

Some techniques used in DX work are not used in any other type of amateur operating. Newcomers find some of them baffling, but study of the situation generally will reveal a purpose for each.

A key reason for certain DX operating procedures used by amateurs in the United States is the fact that *there are so many of us and so few of them*. Amateurs in the United States more than outnumber those in foreign countries, and if sheer numbers is not enough, we also are allowed to use higher power and many of us have the affluence to put up fancier antennas than many foreign hams. So, in amateur radio as in political and social matters, the United States is the "big noise" that makes itself heard all over the globe.

Being heard is easy, but it isn't enough to run up a big score in DX work. That requires two-way contacts, so the specialized skills and techniques of DX work are aimed at achieving these two-way exchanges by being able to receive foreign signals as well as by being able to overpower the competition in trying to talk to them.

One such technique is transmitting and receiving on different frequencies. Ordinarily, this is poor radio practice for most amateur work, but there are several reasons why it is the accepted practice in chasing DX.

One important reason is related to the fact that many stations are often simultaneously calling a specific DX station they have heard. This noisy and regrettable situation, called a *pileup,* is a fact of life that is accepted by DXers as a necessary evil because of the keen competition for DX contacts. If such a pileup occurs on the frequency used by the DX station, then no one can hear the DX station! So, most DX chasers call the DX station a few kHz away from his frequency, and most DX stations, at least experienced ones, will not answer calls made on their own frequency in order to discourage pileups on their own frequency.

Phone operating is the more difficult way to work DX because phone signals occupy so much more band space than cw signals and consequently create more crowding. However, some foreign countries are allowed to work phone on frequencies where cw only is permitted for U. S. hams. This means they can operate in an area of the band where U. S. phone signals cannot clobber them and many DX stations take advantage of this situation to make contacts that would otherwise be impossible.

Incidentally, the fact that some foreign amateur phone work is permitted on frequencies denied to U. S. phone operators is often viewed by foreign hams as one of their few advantages over the frequently high-powered United States stations. Consequently, some foreign hams are opposed to any expansion of U. S. phone bands, because it would rob them of their present opportunity to talk among themselves without being clobbered by strong phone signals from the United States. This side of the coin is often ignored by hams clamoring for the FCC to expand the phone bands. It also explains why some ham groups in the United States oppose expansion of our phone bands, even though this causes them to be criticized as being interested only in cw. The real issue involves the standing of United States amateur activities in the eyes of amateurs and their governments in other parts of the world.

Finding just what frequency the DX station is listening on is sometimes difficult, but it is part of the specialized skill which a DX chaser develops. Sometimes you will hear a DX station call "CQ" or "QRZ?" and then indicate he is listening a certain number of kHz up or down from his own frequency. On cw, he might say just "U5" for "up five," or "D7" for "down seven."

THE RIGHT TIME TO CALL

If you can hear the stateside station (or other DX station) working the DX station you are after, you know the DX station is listening on that frequency, so you can get your transmitter spotted there and be ready to call the DX station *after* he finishes the contact which he has already made.

DX chasing, like contest operating, sounds like dog-eat-dog sometimes, but rabid DXers stoutly maintain there are rules of good conduct in working DX and that one of the main ones is to never call a DX station on top of someone already working him. To be able to conform to this rule, you need to be able to quickly retune your receiver from the stateside station's frequency to the DX station's frequency (or use two receivers, one tuned to each frequency), or else pay very close attention to what the DX station says that will indicate he is completing his contact.

It is tricky business to time your call so he will hear it without starting a pileup that covers your own call, because when you begin calling the DX station, a lot of other DX chasers will hear you and begin calling him too! If you start your call too early, the resulting pileup will build up before the DX station stands by for calls, so he may hear nothing but raucous noise. If you call too late, the pileup will already have started before you get on the air.

Seasoned DX chasers develop a sort of sixth sense that enables them to know the precise split second to make a call to be most likely to be heard—at least that is how they explain it *after* they have made such a contact. If their sixth sense lets them down, they shrug, mutter some philosophical incantation under their breath, and plunge back onto the DX battlefield to try again.

A stateside station calling "CQ DX" generally is considered a rank greenhorn in DX circles, since it is felt that DX stations, especially rare ones, will not answer CQ calls, but will only answer stations calling them specifically. DX stations themselves may call CQ however, and this is one of the things to listen for when hunting DX.

Once a DX station makes that first contact, though, he probably won't call CQ again, because so many other stations will be calling him that he won't need to.

So, *listening* carefully for DX stations or for stations that are talking to DX stations is the main form of exercise utilized by successful DX chasers, rather than pointlessly cluttering the band with a flood of CQ DX calls.

The successful DX hunter develops very skillful ears and uses them for long periods before he flips his transmitter on the air the first time. He listens for the faint, fluttery signals that characterize some DX stations (but not all of them), and he listens to determine their operating habits so he can get himself into the best position and use the best technique to get the DX station to recognize him.

First, he finds out where the DX station is *listening,* as already explained. Then, he listens to see if the DX station is ragchewing, or is making brief snappy contacts, which often is the case. He then tailors his transmitting style to fit what the DX station seems to want.

Some operators try an end run when a pileup occurs on the frequency to which the DX operator apparently is listening. They transmit on the edge of the pileup with the idea that if the pileup is severe, the DX operator may be unable to hear anyone, so he will tune off to one side slightly, trying to find a call he can understand. If the clever DX chaser is heard, he may be the one who captures the prize! You can see there are many little ploys used by DX chasers to bag their game, just as skillful game hunters employ little tricky maneuvers in stalking their prey.

MAKING PROPAGATION CONDITIONS WORK FOR YOU

Understanding propagation conditions is a necessary tool for the DX chaser, who would otherwise waste a lot of time looking for foreign stations at times when long distance skip just isn't in the cards on a particular band.

During periods of high sunspot numbers, most of the DX is worked on 20, 15, and 10 meters. As sunspots decline, 15 and 10 lose their value to the DX chaser, and 20 generally is good only in the daytime. However, as these higher-frequency bands "fade out" to the DX worker, the lower bands begin getting better. DX becomes much easier to work on 40 during periods of low sunspot numbers, especially at night, and even 80 meters is usable at times.

So, the DX worker keeps abreast of changes in sunspot numbers, which take place over periods of many weeks or months, and plans his future DX working schedules accordingly. Summertime, with longer days in the northern hemisphere, also affects propagation because of the longer periods when the ionosphere is subjected to solar effects. These effects generally benefit DX paths on higher frequencies, whereas wintertime, with shorter days, strengthens the paths on 40 and 80.

Forearmed with this understanding of expected propagation conditions, the DX worker then listens to the activity on various bands to confirm that propagation is occurring as he anticipated, or to discover if there are any unexpected changes which may work to his advantage in making contacts in specific parts of the world.

DXpeditions

Some exotic foreign contacts are "manufactured," in effect, by DX hounds through the medium of "DXpeditions." These temporary foreign stations are established by DX enthusiasts by setting up a station in a country which is recognized by DX authorities and does not have a ham living there, or has very few hams and consequently is extremely hard to work.

Operating locations for these radio expeditions have varied from luxury hotels to a pile of rocks under a palm tree. Such operations are publicized in DX circles in advance, with specific operating times and frequencies usually announced to the DX fraternity, and contacts usually are very rapid and very brief . . . but they do suffice to give a number of rabid DXers "one more country."

DX QSLing

Obtaining QSL cards from some foreign operators is almost a science in itself, but ingenious hams have worked out several ways to get this sometimes-difficult job accomplished. The basic problem usually is simply one of the availability of money for postage. An active foreign station is bound to work many more stateside hams than are stateside hams likely to work foreign stations. So, the burden of the postage bill to send a QSL card to every contact would rapidly get out of hand for the foreign ham.

So, many foreign stations do not send QSL cards direct to stateside hams—they just can't afford the cost of postage required to send a separate card to every stateside station they work. To help ease the postage burden for the foreign hams and still enable them to QSL all their stateside contacts, the ARRL has for many years sponsored in the U.S.A. a number of "QSL Bureaus" that will distribute incoming QSL cards for hams in the United States. Separate bureaus are set up for each call district, and foreign hams are invited to send QSL cards in packages to these bureaus. This way, the foreign station saves up his stateside QSL cards for a period of several weeks or so and sends a bundle to each bureau at a much lower cost than if he mailed each card individually to each station he had worked.

These bureaus sort the cards and then distribute them to individual amateurs in their area by means of envelopes bearing postage sent to the bureaus by hams in the district interested in DX cards. Addresses of these bureaus are carried in QST, the monthly magazine published by the ARRL. Any stateside amateur who has worked DX or plans to do so should send an envelope to his bureau to be kept on file until there are a number of cards there addressed to him. When this happens, the envelope is mailed to him, and he is expected to send in another envelope to collect the next batch of cards that arrives at the bureau from foreign stations.

Some foreign amateur groups provide a similar service for hams in their country. Addresses of these foreign QSL bureaus are listed in the *Call Book,* a quarterly publication listing names and addresses of all amateur radio operators in the world.

If direct QSLs are desired from foreign stations, it is a thoughtful courtesy to send the foreign station postage to pay for this special treatment. Sending stamps won't help, because United States postage stamps are worthless in foreign countries. The way to surmount this hurdle is to send *International Reply Coupons,* which can be purchased at United States post offices in various amounts and can be exchanged at foreign post offiices for stamps of that country.

Chapter 10

Operating Across the Border

Amateur radio operating is generally thought of as an activity taking place only at home, but it is becoming increasingly common for an amateur to take his radio operating interests with him on trips, not only throughout his own country, but also into foreign countries and international areas; that is, any area outside the jurisdiction of any national government.

Requirements for operating portable and mobile by amateurs within the United States were covered in Chapter 2. However, additional regulations and requirements exist for operation by United States amateurs outside their own country, and for operation by alien amateurs within the United States.

INTERNATIONAL OPERATING

When an amateur from the United States travels to a foreign country, he may operate in that country if given permission by that country's government. However, in traveling in international territory; an amateur from the United States need only give notice to the FCC, and comply with established regulations for international operating. The requirements for this notice to the FCC are the same as the requirements for notification when operating mobile or portable within the United States away from the home location for more than 48 hours.

The frequencies which amateurs are allowed to use in international territory vary, depending on the part of the world in which the operating occurs. The world is divided into three *regions* by the International Telecommunications Union, which administers worldwide radio regulations. Roughly speaking, Region 1 includes Europe and Africa, Region 2 includes North and South America, and Region 3 includes everything else.

When operating in Region 2, but outside the jurisdiction of a particular country's government, an amateur may operate on any amateur frequency from 7 MHz through 148 MHz. In Regions 1 and 3, when outside the jurisdiction of a country's government, an amateur may operate only from 7 to 7.1 MHz, 14.0 to 14.35 MHz, 21.0 to 21.45 MHz, and 28.0 to 29.7 MHz.

This operation may be at a fixed location, or may be on a ship, aircraft, or land vehicle. When identifying while in international territory, the amateur uses his regular call sign followed by indication of the region in which he is operating. If the operation is on cw, use a slant bar followed by R1, R2, or R3. If on phone, use the call sign, followed by the word "mobile," followed by the words "region one," "region two," or "region three."

So, on cw, typical identification would be "W7JCH/R1," and on phone it would be "W7JCH mobile region one."

TRAVELING IN CANADA

For years, amateurs of the United States have been permitted to operate while traveling in Canada, merely by filing a request for authorization in advance with the Canadian Department of Communication, which is the Canadian radio licensing authority.

A simple post card form is available for doing this and may be obtained from the Canadian Department of Communications, Telecommunications Regulation Branch, 100 Metcalfe St., Ottawa, Canada.

It is also necessary to notify the FCC office which has jurisdiction over your home location. There are no special FCC forms for this; it may be done by letter or post card.

RECIPROCAL LICENSING

United States amateurs also may operate in other foreign countries, and amateurs from those countries may operate in the United States under terms of what is commonly called "reciprocal licensing." Special FCC regulations cover this type of operation, and formal treaties exist between the United States and certain foreign countries to form the necessary international agreements.

As this is written, the countries with which the United States has reciprocal agreements are Argentina, Australia, Austria, Barbados, Belgium, Bolivia, Brazil, Canada, Chile, Colombia, Costa Rica, Dominican Republic, Ecuador, El Salvador, Finland, France, Germany, Guatemala, Guyana, Honduras, India, Indonesia, Ireland, Israel, Jamaica, Kuwait, Luxembourg, Monaco, Netherlands, New Zealand, Nicaragua, Norway, Panama, Paraguay, Peru, Portugal, Sierra Leone, Sweden, Switzerland, Trinidad, Tobago, the United Kingdom, Uruguay, and Venezuela.

In the cases of France, Netherlands, and the United Kingdom, the agreements cover overseas entities of those countries.

Canada has reciprocal agreements with Belgium, Bermuda, Dominican Republic, France, Germany, India, Israel, Luxembourg, Mexico, Netherlands, Nicaragua, Norway, Peru, Portugal, Senegal, Sweden, Switzerland, United Kingdom, U.S., Uruguay, and Venezuela.

These lists are changed occasionally as new countries sign such agreements with the United States or Canada.

If an amateur in the United States or Canada wants to travel to any of these countries and operate a ham station there, he must apply for a permit from the licensing authority of that country. Amateurs from these countries coming to the U. S. or Canada would apply to the FCC or DOC.

FCC rules provide for such permits to be issued for one year at a time, and the FCC must be notified of the alien's location when operating. The alien amateur uses his own foreign call sign, followed by the number of the U. S. call sign district in which he is operating in the same way as a U. S. amateur would do while operating portable or mobile. That is, he would use a slant bar on cw followed by the number, or on phone would use the words "fixed," "portable," or "mobile" and the number. Also, at least once during each contact with another amateur station, the alien amateur must state the geographical location of his station.

Operating permits issued by the FCC to aliens give them full use of all United States amateur bands, regardless of the type of license they hold in their own country. In other words, alien amateurs operating with FCC permission in the United States do not need to abide by the incentive licensing restrictions which apply to all U. S. citizens holding amateur licenses.

BANNED COUNTRIES

A few countries of the world have notified the International Telecommunications Union in past years that they do not allow their own amateurs to communicate with amateurs of other countries. The United States and Canada, members of the ITU, in turn issue their own regulations prohibiting their amateurs from working amateurs in specific countries.

Countries with which U. S. amateurs are forbidden to communicate are Cambodia and Vietnam, which have the call sign prefixes XU, XV, and 3W8. Canadian amateurs are forbidden to contact Cambodia, Vietnam, and Jordan which have the call sign prefixes JY, XU, XV, and 3W8.

As with other international restrictions, the list of banned countries changes from time to time, so it is necessary for U.S. and Canadian amateurs to check the latest FCC or DOC regulations to be sure of their present status.

Appendix

PHONETIC ALPHABETS

	ARRL	ICAO
A	Adam	Alfa
B	Baker	Bravo
C	Charlie	Charlie
D	David	Delta
E	Edward	Echo
F	Frank	Foxtrot
G	George	Golf
H	Henry	Hotel
I	Ida	India
J	John	Juliett
K	King	Kilo
L	Lewis	Lima
M	Mary	Mike
N	Nancy	November
O	Otto	Oscar
P	Peter	Papa
Q	Queen	Quebec
R	Robert	Romeo
S	Susan	Sierra
T	Thomas	Tango
U	Union	Uniform
V	Victor	Victor
W	William	Whiskey
X	X-Ray	X-Ray
Y	Young	Yankee
Z	Zebra	Zulu

RST SIGNAL REPORTING SYSTEM

Readability

1—Unreadable
2—Barely readable, some words distinguishable
3—Readable with considerable difficulty
4—Readable with practically no difficulty
5—Perfectly readable

Signal Strength

1—Faint signal, barely perceptible
2—Very weak signal
3—Weak signal
4—Fair signal
5—Fairly good signal
6—Good signal
7—Moderately strong signal
8—Strong signal
9—Extremely strong signal

Tone

1—Extremely rough hissing note
2—Very rough ac note, no trace of musicality
3—Rough low-pitched ac note, slightly musical
4—Rather rough ac note, moderately musical
5—Musically modulated note
6—Modulated note, slight trace of whistle
7—Near dc note, smooth ripple
8—Good dc note, just a trace of ripple
9—Purest dc note
C—Chirp
K—Key clicks
X—Stability characteristic of crystal control

Q SIGNALS

(To send question form, follow Q signal with question mark.)

For General Operating

QRG What is my frequency in kHz?
Your frequency is_________kHz.

QRH Does my frequency vary?
Your frequency varies.

QRI What is the tone of my signal?
The tone of your signal is_________(1 = good; 2 = variable; 3 = bad).

QRK What is the intelligibility of my signals (or those of_________)?
The intelligibility of your signals (or those of_________) is_________(1 = bad; 2 = poor; 3 = fair; 4 = good; 5 = excellent).

QRL Are you busy?
I am busy (or busy with_________). Please do not interfere.

QRM Is my transmission being interfered with?
Your transmission is being interfered with—(1 = nil; 2 = slightly; 3 = moderately; 4 = severely; 5 = extremely).

QRN Are you troubled with static?
I am troubled with static—(1 to 5, same as under QRM).

QRO Should I increase transmitter power?
Increase transmitter power.

QRP Should I decrease transmitter power?
Decrease transmitter power.

QRQ Shall I send faster?
Send faster (_________words a minute).

QRRR Official ARRL land distress signal. For use in emergency only.

QRS Shall I send slower?
Send slower (_________words a minute).

QRT Shall I stop sending?
Stop sending.

QRW Shall I advise_________that you are calling him on_________kHz?

Advise ________ that I am calling him on ________ kHz.

QRX When will you call again?
I will call you again at ________ hours on ________ kHz.

QRZ Who is calling me?
You are being called by ________.

QSA What is my signal strength (or the strength of ________)?
Your signal strength (or the strength of ________) is ________. (1 = scarcely perceptible; 2 = weak; 3 = fairly weak; 4 = good; 5 = very good).

QSB Is my signal fading?
Your signal is fading.

QSD Are my signals mutilated?
Your signals are mutilated.

QSN Did you hear me (or ________) on ________ kHz?
I heard you (or ________) on ________ kHz.

QSO Can you communicate with ________ direct or by relay?
I can communicate with ________ direct or by relay.

QST Calling all radio amateurs. (Used as a general call preceding transmission of information from ARRL Headquarters by ARRL station W1AW and by Official Bulletin Stations appointed to serve this information function.)

QSV Shall I send a series of Vs?
Send a series of Vs.

QSW Will you send on this frequency (or on ________ kHz)?
I will send on this frequency (or on ________ kHz).

QSX Will you listen for ________ on ________ kHz?
I am listening for ________ on ________ kHz.

QSY Shall I move to ________ kHz?
Move to ________ kHz.

QTH	What is your location? My location is________.
QTR	What is the correct time? The correct time is________.

For Message Handling

QRU	Do you have anything for me? I have nothing for you.
QRV	Are you ready (to receive message)? I am ready.
QRY	What is my turn? Your turn is________.
QSG	Shall I send________messages at a time? Send________messages at a time.
QSK	Can you hear me between the keying of your signal, and if so can I break in on your transmission? I can hear you between the keying of my signal; break in on my transmission when necessary.
QSL	Do you acknowledge receipt (of the message or information)? I acknowledge receipt (of your message or information).
QSM	Shall I repeat the last message I sent you (or some previous message)? Repeat the last message you sent me (or message(s) number(s)________).
QSP	Will you relay to________? I will relay to________.
QSU	Shall I send or reply on this frequency (or on ________kHz)? Send or reply on this frequency (or on________ kHz).
QSZ	Shall I send each word or group twice? Send each word or group twice.
QTA	Shall I cancel message number________? Cancel message number________.

QTB	Do you agree with my word count? I do not agree with your word count. I shall repeat the first letter of each word and the first figure of each number.
QTC	How many messages do you have to send? I have______ messages for you (or for______).
QTX	Will you keep your station open for further communication with me until further notice (or until ______ hours)? I will keep my station open for further communication with you until further notice (or until______ hours).

Special QN Signals for Amateur Radio Net Use

(These carry different meanings from commercial QN signals)

QNA*	Answer in prearranged order.
QNB*	Act as relay between______ and______.
QNC	All net stations copy. I have a message for all net stations.
QND*	Net is directed (controlled by net control station).
QNE*	All net members stand by.
QNF	Net is free (not controlled). Net is closed.
QNG	Take over as net control station.
QNH	Your net frequency is high.
QNI	Net stations report in.* I am reporting into net. (Follow with list of traffic, or QRU.)
QNJ	Can you copy me? Can you copy______?
QNK*	Send messages for______ to______.
QNL	Your net frequency is low.
QNM*	You are interfering with the net. Stand by.
QNN	What station is net control? Net control station is______.

QNO Station is leaving the net.

QNP I am unable to copy you.

QNQ* I am unable to copy__________.
Move to__________kHz and wait for__________to finish handling traffic. Then send him traffic for __________.

QNR* Answer__________and receive traffic.

QNS Following stations are in the net.*
What stations are in the net?

QNT I request permission to leave the net for__________ minutes.

QNU* The net has traffic for you. Stand by.

QNV* Make contact with__________on this frequency. Then move to__________and send him traffic for __________.

QNW How do I route messages for__________?

QNX You are excused from the net.*
Request to be excused from the net.

QNY* Shift to another frequency (or to__________kHz) and clear traffic with__________.

QNZ Zero beat your signal with mine.

*For use only by Net Control Station

Signals for RTTY Use

QJA Is my RTTY (1 = tape; 2 = mark-space) reversed?
It is reversed.

QJB Shall I use (1 = TTY, 2 = reperf)?
Use (1 = TTY; 2 = reperf).

QJC Check your RTTY (1 = TC; 2 = auto head; 3 = reperf; 5 = printer; 7 = keyboard).

QJD Shall I transmit (1 = letters; 2 = figures)?
Transmit (1 = letters; 2 = figures).

QJE Shall I send (1 = wide; 2 = narrow; 3 = correct) shift?
Your shift is (1 = wide; 2 = narrow; 3 = correct).

QJF	Does my signal check out OK? Your signal checks out OK.
QJH	Shall I transmit (1 = test tape; 2 = test sentence)? Transmit (1 = test tape; 2 = test sentence).
QJI	Shall I transmit continuous (1 = mark; 2 = space) signal? Transmit continuous (1 = mark; 2 = space) signal.
QJK	Are you receiving continuous (1 = mark; 2 = space; 3 = mark bias; 4 = space bias)? I am receiving continuous (1 = mark; 2 = space; 3 = mark bias; 4 = space bias).

PROWORDS

(Procedure terminology used in phone operating)

Affirmative	Yes
All after	Say again all your transmission after the words________.
All before	Say again all your transmission before the words________.
Break	First meaning: I wish to break into your transmission. Second meaning: Used to separate the parts of a formal message (preamble, address, text, and signature).
Clear	I am finished transmitting, but will stand by for any calls.
Closing station	The station is being switched off. No other calls will be answered at this time.
Go ahead	Start transmitting
Out	I am ending my transmission and will not be available for any calls.
Over	Start transmitting.
Roger	I understood your entire transmission.
Stand by	Listen, but do not transmit.
Wilco	I will comply.

PROSIGNS

(Procedure signals used on cw)
(Overline indicates letters are run together in sending as if one character)

$\overline{\text{AR}}$	End of message, or end of transmission. Also used after call to a station with which contact has not yet been made, instead of "K."
$\overline{\text{AS}}$	Wait; stand by
$\overline{\text{BT}}$	Used to separate thoughts in casual conversation on CW, where period is not ordinarily used; also used to separate preamble and signature from text of formal message.
C	Yes
CL	Closing station; going off the air for the time being.
DE	From
K	Go ahead
$\overline{\text{KN}}$	Go ahead specific station, or specific area, all other stations stand by.
N	No
R	I understood your entire transmission.
$\overline{\text{SK}}$	End of contract; I am through sending, but will listen for your "final."

COMMON CW ABBREVIATIONS

ABT	About
ADR	Address
AGN	Again
ANT	Antenna
ASAP	As soon as possible
AVG	Average
BC	Broadcast
BCNU	Be seeing you
BK	Break

BTR	Better
CANS	Headphones
CFM	Confirm
CK	Check
CLR	Clear
CNDX	Conditions
CP	Code proficiency
CPI, CPY	Copy
CU	See you
CUAGN	See you again
CUD	Could
CUL	See you later
CUZ	Cause
D, DWN	Down
DLD	Delivered
ENUF	Enough
ES	And
ETA	Estimated time of arrival
ETD	Estimated time of departure
FAX	Facsimile
FB	Fine business
FD	Field day
FER	For
FM, FRM	From
FWD	Forward
GA	Go ahead, good afternoon
GE	Good evening
GESS	Guess
GM	Good morning
GN	Good night
GND	Ground
GUD	Good
HI	Laughter

HPE	Hope
HR	Here
HV	Have
HW	How
ID	Identification
IRC	International Reply Coupon
LID	Poor, discourteous operator
LIL	Little
LTR	Letter
MAX	Maximum
MILL	Typewriter
MIN	Minute; minimum
NIL	Nothing; none
NR	Number; near
NW	Now
OM	Old man (any male ham)
OP, OPR	Operator
OT	Old timer
PD	Period
PLS, PSE	Please
PTT	Push to talk
PWR	Power
RCVR, RX	Receiver
RITE	Right; write
RPT	Repeat; report
RT	Right; route
SASE	Self-addressed stamped envelope
SHUD	Should
SIG	Signature; signal
SKED	Schedule
SRI	Sorry
STN	Station
SUM	Some

TFC	Traffic
TMW	Tomorrow
TNX	Thanks
TU	Thank you
TX	Transmitter
UR	Your; you're
VY	Very
WAT	What
WATSA?	What say?
W, WID	With
WO	Without
WRD	Word
WL	Will; well
WX	Weather
XCVR	Transceiver
XMT	Transmit
XMTR	Transmitter
XTAL	Crystal
XYL	Wife (ex-young lady)
YL	Young lady (often female ham of any age)
YR	Year; your; you're
73	Best regards (*NEVER* sent as "73s" by a good operator!)

PHILLIPS CODE

(Originated as a form of shorthand for telegraph operators nearly 100 years ago. Still useful occasionally by top-flight operators as a time saver. Code comprised of one- and two-letter abbreviations for "root" words, with additional letters indicating other forms of the word. For example, adding "D" forms the past tense of the root word; adding "S" forms the plural; adding "G" forms the past participle; adding "Y" forms the adverb. Other letters are used for other or special forms of root words. Following is a brief list of Phillips Code words to show how the abbreviations are formed.)

AB	About	EFI	Efficient
ABB	Abbreviate	EFY	Efficiency
ABBD	Abbreviated	EMJY	Emergency
ABBG	Abbreviating	ENGR	Engineer
ABBN	Abbreviation	ENGRD	Engineered
ABC	Absence	ENGRG	Engineering
ADVC	Advance	GV	Give
ADVCD	Advanced	GVG	Giving
ADVCG	Advancing	GVN	Given
AMPF	Amplify	HRD	Heard
AMPFD	Amplified	HRG	Hearing
AMPFN	Amplification	HRY	Hurry
AMPFR	Amplifier	HV	Have
BCA	Became	HVB	Have been
BCM	Become	HVNB	Have not been
BF	Before	INC	Increase
BFH	Beforehand	INT	Interest
CAL	Calculate	KPG	Keeping
CALD	Calculated	LAF	Laugh
CALG	Calculating	LTR	Letter
CALN	Calculation	M	More
CFL	Confidential	MAB	May be
CFLY	Confidentially	MDL	Middle
CFY	Chiefly	MDM	Medium
CLR	Clear	MOF	Matter of fact
CLRD	Cleared	MOL	More or less
CLRG	Clearing	NU	New
DL	Deliver	NX	Next
DLD	Delivered	O	Of
DLG	Delivering	OFR	Offer
DLY	Delivery	P	Per

PC	Per cent
PNT	Point
PPN	Proposition
RMX	Remarks
RTN	Return
RTD	Returned
RTG	Returning
RU	Are you
SDN	Sudden
SN	Soon
SND	Send
STY	Steady
SUG	Suggest
SVL	Several
SXH	Sixth
T	The
TCY	Technically
TCL	Technical
TMT	Transmit
TMD	Transmitted
TMG	Transmitting
TMN	Transmission
TR	There
TT	That
TTI	That is
TTT	That the
U	You
UC	You see
UKN	Unknown
VF	Verify
W	With
WD	Would
WH	Which
WO	Who
WR	Were
WY	Why
XC	Excite
Y	Year

10 CODE

(Note: Where used by amateurs operating on f-m in the 6- and 2-meter bands and the 450-MHz band, the exact meanings vary from one locality to another. The codes shown are typical of those sometimes used by some f-m hams, but usage has not become extensive enough to bring about an officially recognized list.)

10-1	Receive you poorly
10-2	Receive you well
10-4	General-purpose acknowledgment
10-6	Busy
10-7	Leaving the air
10-8	Returning to the air
10-9	Repeat
10-10	Will remain on frequency

10-19	Destination
10-20	Location
10-21	Telephone
10-23	Stand by
10-33	Emergency traffic
10-97	Arrived at destination
10-98	Assignment completed

Code 7 (*not* 10-7) sometimes used to mean "out to lunch."

ARRL NUMBERED RADIOGRAMS

For Emergency Situations

ONE	All safe. Do not be concerned about disaster reports.
TWO	Coming home as soon as possible.
THREE	Am perfectly all right. Don't worry.
FOUR	Everyone safe here. Only slight property damage.
SIX	Everyone safe, writing soon.
SEVEN*	Reply by amateur radio.
ELEVEN	Conditions do not permit me to come home at this time. Am perfectly all right. Will be home as soon as possible.
TWELVE*	Are you safe? Anxious to hear from you.
THIRTEEN*	Is________safe? Please advise.
FIFTEEN*	Advise if you need help.
SIXTEEN*	Please advise your condition.
EIGHTEEN*	Please contact me as soon as possible (at________).
NINETEEN*	Request health and welfare report on ________(name, address, telephone number).

Thanks or Social Messages

TWENTY	Thanks for the wonderful time.
TWENTY-ONE	Your gift greatly appreciated. Many thanks.
TWENTY-TWO	Many thanks for your good wishes. Happy to hear from you.
TWENTY-THREE	Your amateur radiogram appreciated. Many thanks.
TWENTY-FOUR	Your letter appreciated. Many thanks.
TWENTY-FIVE	Your package received. Many thanks.
TWENTY-SIX	Delighted to hear the good news.
TWENTY-SEVEN	Congratulations on your promotion.
TWENTY-EIGHT	Congratulations on your election.
THIRTY	Good luck in your new position.

Anniversaries

THIRTY-ONE	Heartiest congratulations on your wedding anniversary.
THIRTY-TWO	Love and best wishes on the anniversary.
THIRTY-FOUR	Wish I could be with you.
THIRTY-FIVE	Wish we could be with you.
THIRTY-SIX	Wish you could be with us.
THIRTY-EIGHT	Congratulations and best wishes on your retirement.

Weddings and Birthdays

FORTY	Congratulations and best wishes for health, happiness and prosperity.
FORTY-ONE	Hearty congratulations and wishes for your happiness together.
FORTY-SIX	Greetings on your birthday and best wishes for many more to come.
FORTY-EIGHT	Congratulations on the new arrival. Hope mother and child are well.

*Not recommended for use into disaster areas. Messages coming out of disasters must receive precedence.

Greetings and Seasonal Messages

FIFTY	Greetings by amateur radio.
FIFTY-TWO	Love and best wishes. I am thinking of you most affectionately on this day.
FIFTY-THREE	Best wishes for a Thanksgiving Day full of cheer and happiness.
FIFTY-FOUR	Victory or defeat, our best wishes are with you. Hope you win.
FIFTY-SIX	All the best for a Merry Christmas.
FIFTY-EIGHT	Wishing you a very Merry Christmas and a Happy New Year.
FIFTY-NINE	Merry Christmas from all of us. Wish you could be here.
SIXTY	All the best in the New Year.
SIXTY-ONE	Love and best wishes for a Merry Christmas and a Happy New Year.
SIXTY-TWO	The heartiest of season's greetings.
SIXTY-THREE	Most sincere wishes for health, happiness and prosperity.
SIXTY-SEVEN	Most sincere wishes for health and happiness on this Easter Day.
SEVENTY	Best wishes for a very pleasant journey.
SEVENTY-ONE	Have not heard from you in some time. Please write or answer by amateur radio, through the station delivering this message.
SEVENTY-TWO	Wishing you and yours the best this holiday season.
SEVENTY-FOUR	Hope you have a wonderful vacation. See you when you get back.
SEVENTY-FIVE	Having a wonderful time. See you when I return.
SEVENTY-SIX	This message is sent free of charge by amateur radio from an exhibit at

	the (name of fair, exposition, hobby show, etc.)
SEVENTY-SEVEN	Having a wonderful time at (name of fair, exposition, exhibit, etc.) Wish you were here.

Miscellaneous

SEVENTY-EIGHT	This acknowledges receipt of your letter.
SEVENTY-NINE	This acknowledges receipt of your message.
EIGHTY	This acknowledges receipt of your recent communication.
EIGHTY-ONE	Please acknowledge receipt of this message by return radiogram.
EIGHTY-TWO	Sorry to hear you are ill. Best wishes for a speedy recovery.
EIGHTY-SIX	Everything well here. Hope all well there. Will write later.
EIGHTY-SEVEN	Send ________ dollars as soon as possible.
EIGHTY-NINE	Am sending ________ dollars immediately.
NINETY	Arrived at ________ safely. Will write as soon as possible.
NINETY-ONE	I am in the hospital and receiving excellent care. Will write soon. Don't worry. My new address is ________________.
NINETY-TWO	Please arrange quarters for ______ days.
NINETY-THREE	Have been reassigned to ________. Will write as soon as possible.
NINETY-FOUR	Will travel on temporary orders to ________ (place) on ______ (date).
NINETY-FIVE	Will arrive ________(place) on ______(date).

NINETY-SIX	Send no further mail. Will send new address soon.
NINETY-SEVEN	Leave granted. Should arrive on ______ ______(date).
NINETY-EIGHT	Leave denied. Will write later.
NINETY-NINE	Please arrange transportation to meet me.

MILITARY ABBREVIATIONS

(Often used in addresses of messages sent to military personnel.)

CO	Company
BN, BTN	Battalion
BRIG, BRD, BDE	Brigade
RGT, REGT	Regiment
PLT, PLAT	Platoon
INF	Infantry
MCZ, MECH	Mechanized
MARDIV	Marine Division
MAW	Marine Air Wing
SDN, SQDN, SQ, SGD	Squadron
GP	Group
SV, SVC	Service
CAV	Cavalry
A/M	Air Mobile
A/B, AB	Airborne
HQ	Headquarters
HV	Heavy

INTERNATIONAL RESTRICTIONS

Prohibited Countries

Cambodia and Vietnam forbid radio communications between their amateur stations and amateurs in other countries. The forbidden prefixes for United States amateurs are XU, XV, and 3W8.

Canadian amateurs may not communicate with Jordan, in addition to Cambodia and Vietnam. Forbidden prefixes for contacts by Canadian amateurs are JY, XU, XV, and 3W8.

Third Party Restrictions

Messages on behalf of nonamateurs may be handled by U.S. amateurs only with amateurs in these countries: Argentina, Barbados (only U.S. stations/8P), Bolivia, Brazil, Canada, Chile, Colombia, Costa Rica, Cuba, Dominican Republic, Ecuador, El Salvador, Greenland (XP calls only), Haiti, Honduras, Israel, Liberia, Mexico, Nicaragua, Panama, Paraguay, Peru, Uruguay, and Venezuela. Permissible prefixes are: CE, CM, CO, CP, CX, EL, HC, HH, HI, HK, HP, HR, LU, OA, PY, TI, VE, VO, W or K/8P, XE, XP, YN, YS, YV, ZP, 4X, and 4Z.

Canadian amateurs may handle messages on behalf of nonamateurs only with amateurs in these countries: Bolivia, Chile, Costa Rica, Dominican Republic, El Salvador, Honduras, Israel, Mexico, Peru, U.S., and Venezuela. Permissible prefixes are: CE, CP, HI, HR, K, OA, TI, W, XE, YS, YV, 4X and 4Z.

Reciprocal Operating Agreements

Reciprocal operating agreements exist between the United States and these countries: Argentina, Australia, Austria, Barbados, Belgium, Bolivia, Brazil, Canada, Chile, Colombia, Costa Rica, Dominican Republic, Ecuador, El Salvador, Finland, France*, Germany, Guatemala, Guyana, Honduras, India, Indonesia, Ireland, Israel, Jamaica, Kuwait, Luxembourg, Monaco, Netherlands*, New Zealand, Nicaragua, Norway, Panama, Paraguay, Peru, Portugal, Sierra Leone, Sweden, Switzerland, Trinidad and Tobago, United Kingdom*, Uruguay, and Venezuela. Some other foreign countries grant amateur operating privileges to FCC licensees on a courtesy basis. Details are available from ARRL headquarters.

Reciprocal operating agreements exist between Canada and Belgium, Bermuda, Dominican Republic, France, Germany, India, Israel, Luxembourg, Mexico, Netherlands, Nicaragua, Norway, Peru, Portugal, Senegal, Sweden, Switzerland, United Kingdom, U.S., Uruguay, and Venezuela.

TIME CONVERSION CHART

GMT	EDT/AST	CDT/EST	MDT/CST	PDT/MST	PST
0000	2000	1900	1800	1700	1600
0100	2100	2000	1900	1800	1700
0200	2200	2100	2000	1900	1800
0300	2300	2200	2100	2000	1900
0400	0000	2300	2200	2100	2000
0500	0100	0000	2300	2200	2100
0600	0200	0100	0000	2300	2200
0700	0300	0200	0100	0000	2300
0800	0400	0300	0200	0100	0000
0900	0500	0400	0300	0200	0100
1000	0600	0500	0400	0300	0200
1100	0700	0600	0500	0400	0300
1200	0800	0700	0600	0500	0400
1300	0900	0800	0700	0600	0500
1400	1000	0900	0800	0700	0600
1500	1100	1000	0900	0800	0700
1600	1200	1100	1000	0900	0800
1700	1300	1200	1100	1000	0900
1800	1400	1300	1200	1100	1000
1900	1500	1400	1300	1200	1100
2000	1600	1500	1400	1300	1200
2100	1700	1600	1500	1400	1300
2200	1800	1700	1600	1500	1400
2300	1900	1800	1700	1600	1500
2400	2000	1900	1800	1700	1600

*Agreement includes overseas entities.